T0296278

Artificial Intelligence in Bioinformatics

Artificial Intelligence in Bioinformatics

From Omics Analysis to Deep Learning and Network Mining

Mario Cannataro
University "Magna Græcia" of Catanzaro
Department of Medical and Surgical Sciences and
Data Analytics Research Center
Catanzaro, Italy

Pietro Hiram Guzzi
University "Magna Græcia" of Catanzaro
Department of Medical and Surgical Sciences
Catanzaro, Italy

Giuseppe Agapito
University "Magna Græcia" of Catanzaro
Department of Law, Economy and Sociology
Catanzaro, Italy

Chiara Zucco
University "Magna Græcia" of Catanzaro
Department of Medical and Surgical Sciences
Catanzaro, Italy

Marianna Milano
University "Magna Græcia" of Catanzaro
Department of Medical and Surgical Sciences
Catanzaro, Italy

Elsevier
Radarweg 29, PO Box 211, 1000 AE Amsterdam, Netherlands
The Boulevard, Langford Lane, Kidlington, Oxford OX5 1GB, United Kingdom
50 Hampshire Street, 5th Floor, Cambridge, MA 02139, United States

ISBN: 978-0-12-822952-1

For information on all Elsevier publications
visit our website at https://www.elsevier.com/books-and-journals

Publisher: Oliver Walter
Acquisitions Editor: Priscilla Braglia
Editorial Project Manager: Susan E. Ikeda
Production Project Manager: Kiruthika Govindaraju
Designer: Miles Hitchen

Typeset by VTeX

This book is for my wife Angela, my sons Francesco and Matteo, and my daughter-in-law Andrea, for their interest in the themes of this book, their patience and support.

Mario Cannataro

To Anna, Pietro, Fernando Jr., Fernando Sr., and Salvatore. In particular, to Anna for helping me in losing files, and Fernando for helping me with closed forms.

Pietro Hiram Guzzi

I express my profound gratitude to my parents Divina and Mario who always believed that I could do anything; to my siblings Barbara and Michele for their support and patience. My partner Andrea always has good words for me, radiating a source of strength and inspiration. And very profound gratitude to all the special people who are no longer next to me.

Giuseppe Agapito

To my wonderful family, and to my little Irene, who makes me learn something new every day.

Chiara Zucco

To my parents, my sister, Morghy, and Susi. Their huge love encourages me every moment.

Marianna Milano

Contents

About the authors

Mario Cannataro is a Full Professor of computer engineering and the Director of the Data Analytics Research Center at the University "Magna Græcia" of Catanzaro, Italy. His current research interests include bioinformatics, health informatics, artificial intelligence, data mining, and parallel computing. His main scientific contributions are in the fields of: parallel and distributed data mining and artificial intelligence; problem-solving environments for the analysis of omics data; and efficient bioinformatics and health informatics software tools for the analysis, integration, and management of biosignals, molecular, and clinical data. He has published four books and more than 300 papers in international journals and conference proceedings. Mario Cannataro is Editor-in-Chief of the Encyclopedia of Bioinformatics and Computational Biology and Associate Editor of Briefings in Bioinformatics and the IEEE/ACM Transactions on Computational Biology and Bioinformatics journals. He is a Senior Member of ACM, ACM SIGBio, IEEE, IEEE Computer Society, BITS (Bioinformatics Italian Society), and SIBIM (Italian Society of Biomedical Informatics). He regularly co-organizes international workshops on bioinformatics and high-performance computing in primary conferences such as ACM-BCB, IEEE-BIBM, and ICCS.

Pietro Hiram Guzzi is an Associate Professor of computer engineering at the University "Magna Græcia" of Catanzaro, Italy. His research focuses on developing novel methods for the efficient analysis of biological networks derived from large (big) biomedical data generated from high-throughput experiments. The ultimate goal of the research is to support the development of novel drugs in a translational and precision medicine scenario. From a computational point of view, his research interests include, but are not limited to, network generation and analysis (including comparison and integration), integrating both clinical and geographical data. From a biological point of view, his research focuses on gene and protein function analysis, as well as the identification of interplay among molecules related to disease insurgence and progression. More recently, Pietro is involved in the analysis of diseases integrating heterogeneous data into a single model based on networks and the development of novel methods for brain data analysis. His research efforts have resulted in three books, more than 50 peer-reviewed publication in top-class journals, and more than 70 papers presented at bioinformatics conferences.

Giuseppe Agapito is Assistant Professor of computer engineering at the University "Magna Græcia" of Catanzaro, Italy. He received his PhD in biomedical and computer engineering from the University "Magna Græcia" of Catanzaro, Italy, in 2013. His current research interests include analysis and visualization of biological networks, efficient analysis of genomics data, parallel computing, and data mining. In particular, his research activity is focused on the development and implementation of statistical and data mining methodologies that are also based on parallel and distributed computing for the efficient analysis of omics data. He has published over 70 articles for international journals and conference proceedings. He is a member of the ACM, ACM SIGBio, and BITS. He has also edited a book for Springer.

Chiara Zucco is PostDoc in computer engineering at the University "Magna Græcia" of Catanzaro, Italy. She received her Ph.D. Degree in Biomarkers of Chronic and Complex Diseases at the University "Magna Græcia" of Catanzaro, Italy, in 2020. She has published over 20 articles in international journals and conference proceedings. Her research interests comprise analysis of text for patient monitoring and adverse event prediction.

Marianna Milano is Assistant Professor of computer engineering at the University "Magna Græcia" of Catanzaro, Italy. She received her Master's Degree in Computer Engineering from the University "Magna Græcia" of Catanzaro, Italy, in 2011 and the Ph.D. Degree in Biomarkers of Chronic and Complex Diseases at the University "Magna Græcia" of Catanzaro, Italy, in 2019. She has published over 50 articles in international journals and conference proceedings. Her research interests comprise semantic-based and network-based analysis of biological and clinical data. She is a member of BITS.

Preface

Artificial intelligence is used in several application domains, from robotics to big data analytics, to solve problems with improved accuracy or to add "intelligence" to numerous human processes (e.g., autonomous driving).

Recently, artificial intelligence and its constituent methods, such as machine learning, deep neural networks, data mining, computer vision, and network analysis, to cite a few, have started to be used in bioinformatics, medical informatics, and in several hard sciences, such as physics, chemistry, and biology.

Artificial intelligence is often treated in computer science texts focusing on computational aspects and with limited analysis of bioinformatics applications. On the other hand, bioinformatics is often treated focusing on the basic methods and algorithms and with limited space dedicated to the use of artificial intelligence methods.

To remedy this situation, the book in hand combines a rigorous introduction of artificial intelligence methods in the context of bioinformatics, with a deep and systematic review of how those methods are incorporated in bioinformatics tasks and pipelines.

The book aims to review the main applications of artificial intelligence in bioinformatics, from omics data analysis to deep learning and network mining. The book first recounts the primary methods and theory behind artificial intelligence, including some emergent fields such as sentiment analysis and network alignment. Then it surveys how artificial intelligence is exploited in core bioinformatics applications, including sequence analysis, structure analysis, functional enrichment analysis, protein classification, omics data analysis, integrative bioinformatics, biological networks and pathways analysis, network embedding, ontologies, text mining, and explainable models in bioinformatics. Finally, an overview of the challenges of artificial intelligence in bioinformatics is reported.

Why you should read this book now

With the advent of the omics sciences, the inclusion of molecular data in clinical research, and the considerable production of huge volumes of omics, molecular and clinical data (e.g., lab data and diagnostic images), the bottleneck in bioinformatics analysis has moved from the wet lab to the *in silico* data analysis pipeline, posing novel challenges to bioinformatics, partly solved with the help of parallel computing and cloud infrastructure.

On the other hand, this big data trend in bioinformatics is an exceptional opportunity to empower the bioinformatics pipelines with the power of artificial intelligence. In fact, the availability of huge data at the biological, molecular, tissue, organ, disease, and population levels, is the first step toward the effective training of artificial intelligence algorithms and the realization of biomedical applications according to the bench-to-bedside paradigm.

Because it is possible to foresee that artificial intelligence will more and more be used in bioinformatics, this book is a timely resource in those two fields. In fact, the book describes both the main artificial intelligence methods from a computer science point of view, as well as the role of artificial intelligence in primary bioinformatics methods.

Who should read this book

The book presents the main methods and approaches central to artificial intelligence and bioinformatics and is an important resource for researchers, biologists, and computer scientists working on bioinformatics, molecular biology, and biomedicine.

Readers may find a complete survey of artificial intelligence methods and bioinformatics applications that can guide them in choosing the right artificial intelligence method for improving a specific bioinformatics analysis.

Readers may better understand how the integration of artificial intelligence and bioinformatics methods can enhance the bioinformatics pipelines and the analysis of biological processes.

The book can be used as an introductory book on artificial intelligence and bioinformatics, as well as an advanced research tool for researchers working on emerging bioinformatics fields, such as analysis of omics data, analysis of molecular networks, analysis of molecular pathways, and ontological analysis.

The intended audience of the book comprises researchers and practitioners, as well as post-graduate and PhD students, working on artificial intelligence, bioinformatics, molecular biology, biomedicine, and biotechnology.

The book may also be used as a textbook in artificial intelligence, bioinformatics or computational biology courses at the university level.

A main benefit for the readers is the possibility to have, in a single book, a comprehensive summary of the methods of artificial intelligence and the applications of bioinformatics. On the other hand, the prior knowledge needed to read the book is some basic knowledge of biology and computer science.

How this book is organized

The book comprises 16 chapters organized into two parts: a first, methodological part, that is dedicated to basic methods of artificial intelligence, and a second, applicative part, that is dedicated to classical and emergent applications of bioinformatics. The book is completed by two appendices containing the Java and Python source codes described in the book.

The contents of the 16 chapters and two appendices are summarized below.

Chapter 1 introduces knowledge representation and reasoning, which are key concepts used by artificial intelligence systems to build their knowledge base and to generate new knowledge.

Chapter 2 introduces machine learning methods, which are key concepts used by artificial intelligence systems to learn from data.

Chapter 3 describes the history of artificial intelligence and its relationship with bioinformatics and life sciences.

Chapter 4 introduces data science, a novel discipline that uses the scientific method to combine statistics, data mining, mathematics, and other basic sciences, with the aim of extracting knowledge from data.

Chapter 5 introduces deep learning methods with special focus on artificial neural networks, a key technology that is used by artificial intelligence systems to learn from big data, particularly useful to deal with big bioinformatics data.

Chapter 6 introduces the explainability problem of artificial intelligence methods with a special focus on explainability of machine learning and deep learning methods. Solving the explainability problems is particularly useful in artificial intelligence systems used to make decision in medicine and biology.

Chapter 7 introduces intelligent agents, an abstraction particularly useful to build computational intelligent agent systems that act intelligently and independently in an environment and whose decision can be computationally implemented.

Chapter 8 introduces sequence analysis, a keystone of traditional bioinformatics (sequence alignment algorithms are central to compare DNA or amino acids sequences), which is gaining more interest due to the development of next-generation sequencing that is generating huge volumes of sequence data.

Chapter 9 introduces structure analysis, another keystone of traditional bioinformatics. This chapter focuses on algorithms and methods for the prediction of secondary and tertiary structures of proteins, which have been improved in recent years by the application of artificial intelligence.

Chapter 10 introduces omics sciences, which are having an important role in revolutionizing the bioinformatics world. The chapter describes the most important omics sciences, such as genomics, trascriptomics, epigenomics, proteomics, metabolomics, and relevant data analysis methods.

Chapter 11 introduces ontologies, a computer science approach to build knowledge bases for many domains including bioinformatics. The chapter introduces main biomedical ontologies, including gene ontology and presents the main data analysis methods realizable through ontologies, such as molecular data annotation, semantic similarity measures, and functional enrichment analysis.

Chapter 12 introduces integrative bioinformatics, a novel discipline that aims to improve the analysis of biological data through a systematic integration of several data sources. The chapter discusses the integration and analysis of heterogeneous omics data and describes the most important publicly available data sources that allow researchers to study several diseases by leveraging big data sets of genomics and proteomics data.

Chapter 13 introduces biological networks, a keystone of modern bioinformatics. It describes protein interaction networks and the main methods developed for motif discovery and network comparison.

Chapter 14 introduces biological pathways, a special case of biological networks, that are central to capture the dynamics of biological networks.

Chapter 15 introduces an emerging approach to study biological and clinical data based on text mining, i.e., knowledge extraction from written texts.

Chapter 16 recalls the evolution of bioinformatics and summarizes issues and challenges of using artificial intelligence in bioinformatics that may prevent its use in real applications.

Finally, Appendices A and B report some source codes, respectively in the Python and Java programming languages, that are employed throughout the book to exemplify bioinformatics and artificial intelligence algorithms.

The book offers two reading levels, introductory and advanced. Introductory chapters may be skipped by experienced readers. A set of introductory Chapters 1, 2, 3, 7, 8, and 9 contain introductory material regarding knowledge representation and reasoning, machine learning, artificial intelligence, intelligent agents, and sequence and structure analysis. On the other hand, the remaining chapters provide advanced material regarding data science, deep learning, explainability in artificial intelligence, omics sciences, ontologies, biological networks and pathways analysis, text mining, and the primary challenges and issues of artificial intelligence and bioinformatics.

M. Cannataro, P.H. Guzzi, G. Agapito, C. Zucco, and M. Milano
University "Magna Græcia" of Catanzaro, Catanzaro, Italy
January 31, 2022

Acknowledgments

We thank the University "*Magna Græcia*" of Catanzaro and its School of Informatics and Biomedical Engineering, where we started and consolidated our artificial intelligence and bioinformatics research. Deep appreciation is due to our colleagues working on biology and medicine who presented us many and interesting research problems requiring bioinformatics and artificial intelligence solutions. Special thanks also go to all the faculties, postdocs and PhD students of the Bioinformatics Laboratory for their collaboration on bioinformatics and artificial intelligence research.

We are grateful to Prof. Domenico Talia, Prof. Concettina Guerra, and Prof. Igor Jurisica who introduced us to many of the research topics discussed in this book, such as data mining, artificial intelligence, biological networks, and pathways.

We thank Prof. Tijana Milenkovic and Dr. Marinka Zitnik for a fruitful collaboration in network science.

We also thank Dr. Sergio Bella and Prof. Pierangelo Veltri who collaborated with us on medical and health informatics applications.

Many of the themes described in the book were discussed at some conferences and workshops where we were involved, so special thanks are due to the scientific organizers of the following conferences and workshops: International Conference on Computational Science (ICCS)—Workshop on Biomedical and Bioinformatics Challenges for Computer Science (BBC); ACM International Conference on Bioinformatics, Computational Biology, and Health Informatics (ACM-BCB)—Workshop on Parallel and Cloud-based Bioinformatics and Biomedicine (ParBio); IEEE International Conference on Bioinformatics and Biomedicine (IEEE-BIBM)—International Workshop on High Performance Bioinformatics and Biomedicine (HiBB).

Special thanks also go to the publishing team at Elsevier, whose contributions throughout the whole process from inception of the initial idea to final publication have been invaluable. In particular thanks to Susan Ikeda, who continuously prodded us via e-mail to keeping the project on schedule, and to Priscilla Braglia, who first helped us in this exciting project, and for her guidance throughout the publishing process.

M. Cannataro, P.H. Guzzi, G. Agapito, C. Zucco, and M. Milano
University "Magna Græcia" of Catanzaro, Catanzaro, Italy
January 31, 2022

PART

1

Artificial intelligence: methods

Part 1 outline

Artificial intelligence involves aspects arising from various disciplines, from philosophy to computer science. The first part of the book deals with the basic concepts of artificial intelligence from the computer science point of view, and comprises seven chapters.

Chapter 1 introduces knowledge representation and reasoning, which are key concepts used by artificial intelligence systems to build their knowledge bases and to generate new knowledge.

Chapter 2 introduces machine learning methods, which are key concepts used by artificial intelligence systems to learn from data.

Chapter 3 describes the history of artificial intelligence and its relationship with bioinformatics and life sciences.

Chapter 4 introduces data science, a novel discipline that uses the scientific method to combine statistics, data mining, mathematics, and other basic sciences, with the aim to extract knowledge from data.

Chapter 5 introduces deep learning methods with a special focus on artificial neural networks, a key technology which is used by artificial intelligence systems to learn from big data and is particularly useful to deal with extensive bioinformatics data.

Chapter 6 introduces the explainability problem of artificial intelligence methods, with special focus on the explainability of machine learning and deep learning methods.

Finally, Chapter 7 introduces intelligent agents, an abstraction particularly useful to build computational intelligent agent systems that act intelligently and independently in an environment, whose decisions can be computationally implemented.

CHAPTER

Knowledge representation and reasoning

1

CONTENTS

Abstract

This chapter introduces knowledge representation and reasoning and its relationship with computer systems and artificial intelligence. Knowledge representation and reasoning is a key area of artificial intelligence that involves the (symbolic) representation of knowledge and its automatic manipulation through reasoning programs, with the aim to produce new knowledge or to explain some already established knowledge. A summary of the relationship between artificial intelligence and bioinformatics is also reported.

Keywords

Knowledge representation, Reasoning, Logic, Artificial intelligence, Bioinformatics

1.1 Introduction

Intelligence and intellect are properties of humans beings whose study started in Ancient Greek philosophy. Intellect describes the ability of the human mind to decide what is true and what is false, and to solve problems. It derives from the Ancient Greek term *nous*, that is translated into the Latin term *intellectus* (from *intelligere*, "to understand") and the French and English terms *intelligence*.

Intelligence has many definitions (e.g., logic ability, understanding, self-awareness, learning ability, problem-solving), but in an intuitive way it can be defined as the

Artificial Intelligence in Bioinformatics. https://doi.org/10.1016/B978-0-12-822952-1.00010-3

ability to infer information, to maintain it as knowledge, and to use it when interacting with the environment.

Considering the etymology of the terms intellect and intelligence, intelligence denotes "to gather in between", whereas intellect denotes "what has been gathered", thus intelligence is about the creation of new knowledge, while intellect is about the understanding of existing knowledge.

When discussing human intelligence, a key aspect of intelligent behavior is its relationship with human knowledge. In this context, in intelligent behavior, decisions are taken on the basis of what is known (or believed to be known) about a given domain. Thinking or reasoning, in a very intuitive way, is about how knowledge is elaborated to make decisions or to generate new knowledge.

An agreed definition of Artificial Intelligence (AI) is the study of intelligent behavior achieved through computation. Thus, in the context of AI, knowledge representation and reasoning regard how existing knowledge is represented, stored, and elaborated to decide what to do.

Knowledge Representation and Reasoning (KRR) is a key area of artificial intelligence that involves the (symbolic) representation of knowledge and its automatic manipulation through reasoning programs, with the aim to produce new knowledge or to explain some existing knowledge [1].

A main idea behind knowledge representation and reasoning is the focus on the knowledge, rather than on who holds that knowledge, and then on the mechanisms (reasoning) that enable to discover new knowledge or to react to the environment, on the basis of the already existing knowledge.

A key concept when studying knowledge representation and reasoning is the continuous relationship and interplay between the representation of knowledge and reasoning or, in other words, the description of how we can represent knowledge in the most complete way and how we can reason about it in the most effective and efficient way.

In this context, knowledge is our understanding about the world or a domain, while reasoning is a computational process using such knowledge. To make the entire process efficient and easily presentable, knowledge is represented through a symbolic structure, and reasoning is performed by computational processes that manipulate that symbolic structure.

The rest of the chapter will focus on the main concepts of knowledge representation and reasoning, its main implementations, and its role in building knowledge-based bioinformatics systems.

1.2 Knowledge representation

Although the knowledge concept has been studied since ancient times by Greek philosophers, we will discuss it here in a more informal way. Informally, knowledge is the understanding of someone or something, including facts (also known as descriptive knowledge), skills (also known as procedural knowledge), or objects (also known as acquaintance knowledge) [1,2].

In *descriptive knowledge* (also known as propositional knowledge or declarative knowledge), knowledge can be expressed in a declarative sentence or with an indicative proposition. Usually, it treats explicit knowledge (or expressive knowledge), that is knowledge that can be easily codified, stored, and transmitted. *Procedural knowledge* is about knowing how to perform some task. *Knowledge by acquaintance* refers to familiarity with a person, place, or thing, and concerns perceptual experience (e.g., "I know Mario"). According to Bertrand Russell, acquaintance is a direct causal interaction between a person and some object that the person is perceiving.

An informal way to think about knowledge is to consider knowledge as a relationship between a knower, such as person or an agent, and a proposition that is the fact expressed by a simple declarative sentence. For instance, in the sentence "Mario knows that Rome is the capital of Italy", "Mario" is the knower and "Rome is the capital of Italy" is a proposition.

An important aspect is the nature of propositions. In essence, propositions are abstract entities that can be true or false. In an informal way, representation is a relationship between *two domains*, i.e., we can think that the first domain "stands for" or "takes the place" of the second one. The first domain, called the representor, is usually more concrete than the second one. For example, the image of a circle with an arrow stands for the male abstract symbol.

In KRR the representor is usually a formal symbol, i.e., a character or a group of characters taken from some predetermined alphabet. For instance, the digit "5" stands for the number five, and the letter "V" stands for the fifth element. The use of symbols is motivated by the fact that it is usually easier to work with symbols than with the objects that the symbols represent. This means that it is easier to recognize the symbols, i.e., to distinguish them from each other, etc.

A special case is when a group of symbols stands for a proposition. For instance, "Mario likes bioinformatics" stands for the proposition that Mario likes bioinformatics. In this example, the English sentence has three distinguishable parts (Mario, likes, bioinformatics) and a well-known syntax. On the other hand, the proposition is abstract, and it assumes a classification of all the possible ways in which we can divide the world into two classes: the classes in which Mario likes bioinformatics and those in which he does not.

Knowledge representation thus concerns the use of formal symbols to represent a collection of propositions believed by some agent or person. It should be noted that these symbols do not necessarily represent all the propositions believed by the agent. In fact, there may be an infinite number of propositions believed, but only a finite number of which are represented. It is the role of reasoning to bridge the gap between what is represented and what is believed.

1.3 Reasoning

Reasoning is the formal manipulation of the symbols that represent a collection of believed propositions, with the aim to produce the representations of new propositions [1,2].

From the previous sentence, it emerges that reasoning is about manipulation, and so it is now clear that making manipulation efficient and easy to implement requires that the representation of knowledge, i.e., the symbols, is more accessible than the propositions they represent. Thus, it is the manipulation, i.e., the essence of reasoning, that requires that symbols are concrete enough so that they can be manipulated to build the representations of new propositions.

As an example, if we consider the two propositions "Mario likes bioinformatics" and "Bioinformatics studies computer science and biology", after some manipulation, reasoning could generate the new proposition "Mario likes computer science". This form of reasoning is also known as logical inference because the final proposition is a logical conclusion of the initial ones. In other words, we can regard reasoning as a form of calculation that operates over symbols representing propositions.

1.4 Computer science and knowledge representation and reasoning

Knowledge representation and reasoning aims at building computing systems that, like humans, have some knowledge about the world and behave in an informed way. As noted before, in such systems the knowledge is represented symbolically, and the ability to act is represented by reasoning procedures that can manipulate such stored knowledge and can extract the consequences of such knowledge, that is in turn represented in a symbolic way.

Such an approach has a long history that originated in the field of philosophy, although not yet related to the notion of computation or to the symbolic aspects of knowledge representation; it then continued to evolve, and most recently it has emerged in computer science [3,4].

In fact, first Aristotle developed the initial notion of logic, then Leibniz[1] introduced the notion of "thinking as computation", eventually Frege[2] introduced the

[1] In 1666 Gottfried Wilhelm Leibniz published his dissertation *On the Combinatorial Art* in which he described a theory for the automatic production of knowledge through the rule-based combination of symbols.

[2] Gottlob Frege was a mathematician, logician, and philosopher who worked on the discipline of logic and built a formal system which was the first 'predicate calculus'.

concept of symbolic logic, and lastly Church,[3] Godel,[4] and Turing,[5] connected the basic aspects of symbolic logic with computation and computer science.

A major contribution to the development of artificial intelligence was made by John McCarthy [5], a computer and cognitive scientist who is universally recognized as one of the founders of AI. In fact, he coined the term *artificial intelligence* in the so-called Dartmouth proposal, i.e., the proposal of the "Dartmouth Summer Research Project on Artificial Intelligence" that was co-organized by John McCarthy, Marvin Minsky, Nathaniel Rochester, and Claude Shannon and held in 1956.

The Dartmouth Proposal stated:

"We propose that a 2-month, 10-man study of artificial intelligence be carried out during the summer of 1956 at Dartmouth College in Hanover, New Hampshire. The study is to proceed on the basis of the conjecture that every aspect of learning or any other feature of intelligence can in principle be so precisely described that a machine can be made to simulate it. An attempt will be made to find how to make machines use language, form abstractions and concepts, solve kinds of problems now reserved for humans, and improve themselves. We think that a significant advance can be made in one or more of these problems if a carefully selected group of scientists work on it together for a summer."

Moreover, McCarthy invented Lisp, the family of functional languages based on the lambda calculus, which became the programming language for early AI applications.

McCarthy first conceived the idea that building intelligent systems was enabled by the capability to provide computer systems the knowledge necessary to behave, and not by the details of programming or architecture. If a system is capable to represent what it needs to know (knowledge representation) and it is able to draw conclusions (reasoning) from such knowledge in an automatic way, then it could learn how to behave in an intelligent way.

---

[3] Alonzo Church was a mathematician and logician who worked on mathematical logic and can be considered the founder of theoretical computer science. He defined the lambda calculus, a formalism that allows to define a class of functions that represent the so-called computable functions, i.e., those functions that can be evaluated by means of an algorithm and thus by a computer program. Moreover, in 1936 Alonzo Church and Alan Turing independently published papers showing that a general solution to the *Entscheidungsproblem* (the German term for "decision problem") is impossible. They described the intuitive notion of "effectively calculable" that is represented by the functions computable by a Turing machine or by those representable in the lambda calculus. This concept is also known as the Church–Turing thesis. Alonzo Church and Alan Turing are considered to be the founders of computer science.

[4] Kurt Godel was a logician and mathematician who had a significant impact on scientific and philosophical thinking with a special focus on the use of logic. Godel's main contribution was his first incompleteness theorem and the technique (i.e., Godel numbering which codes formal expressions as natural numbers) that he devised to prove it.

[5] Alan Turing was a mathematician and computer scientist with a central role in the development of theoretical computer science. He formalized the concepts of algorithm and computation with the Turing machine, which constitutes a model of a general-purpose computer. Alan Turing is universally recognized as the father of theoretical computer science and artificial intelligence.

1.5 Artificial intelligence and knowledge representation and reasoning

An important connection between AI and knowledge representation and reasoning was provided by the knowledge representation hypothesis defined by Brian Smith [6]:

"Any mechanically embodied intelligent process will be comprised of structural ingredients that a) we as external observers naturally take to represent a propositional account of the knowledge that the overall process exhibits, and b) independent of such external semantic attribution, play a formal but causal and essential role in engendering the behaviour that manifests that knowledge."

According to this hypothesis, a concrete knowledge representation and reasoning system, called a knowledge-based system, is based on a symbolic representation of knowledge, called the knowledge base (KB).

AI is a discipline of computer science and engineering that aims to create artifacts or programs that exhibit intelligent behavior. Often, the ability of a program to exhibit intelligent behavior is translated as the ability to pass the so-called Turing test [7,8].[6]

Knowledge representation and reasoning is a sub-field of AI concerned with modeling, designing, and implementing the representation of information and knowledge by computers so that programs can use it and, in particular, can reason using it. For instance, an example of reasoning is the derivation of information that is implied by information already stored into the system.

Of course, knowledge representation implementations would be useless without a reasoning system, i.e., that part of the system able to reason with the represented knowledge and its concrete values stored in a knowledge base. Thus, this connection between the knowledge representation and a reasoning system leads to the commonly used term "knowledge representation and reasoning", often denoted as KRR. Often, we say that KRR is necessary and sufficient to produce intelligent behavior.

AI systems based on KRR, other than the ability to derive information that is implied by the knowledge stored in a KB, may have other abilities such as: a) conduct a dialogue with people using a natural language; b) decide what to do, e.g., the next move in a chess game; c) solve problems in domains that usually require human expertise, etc.

Some key concepts useful to characterize a knowledge-based system and thus an AI system are [2]:

[6] In the famous 'Mind paper' of 1950, Alan Turing argues that the question "Can a machine think?" should be replaced with the question "Can a machine be linguistically indistinguishable from a human?" To answer this question, he proposed a test called the 'Turing test'. In this test, a woman and a computer are isolated in separate rooms, and a human judges which rooms contain the woman and the computer, and he questions both the human and the computer by electronic communication (e.g., email; the original term was teletype). If, on the basis of the answers, the judge can do no better than 50/50 when deciding which room houses which player, we say that the computer in question has passed the Turing test. In this sense, passing the test makes linguistic indistinguishability operational.

1. "the things that the system already knows", i.e., a knowledge base of beliefs (e.g., facts) about a certain domain (the world);
2. "the system has to be able to be updated with new knowledge", i.e., there exists a way to add additional beliefs to the knowledge base;
3. "the system must be able to automatically deduce by itself a large set of immediate consequences", i.e., there exists a reasoning method able to generate new beliefs on the basis of the beliefs already contained in the knowledge base.

1.6 Languages for knowledge representation and reasoning

As stated before, especially in the philosophy field, several formal systems have been introduced with the aim to implement human reasoning through automatized computation. The first formal systems were all based on logic that, informally speaking, can be thought of as the study of correct reasoning [2].

Since logic is relevant to knowledge representation and reasoning, in fact it is related to the study of truth conditions and rules of inference. For example, the predicate calculus, also known as the language of first-order logic (FOL), was one of the first logic languages used in AI and KRR, although it was originally devised by Gottlob Frege for formalizing the mathematical inference. On the other hand, there exist several other logic languages for use in AI.

A logic system comprises two main parts: the logic language that is characterized by its syntax and semantics and the method of reasoning [2]. *Syntax* regards the set of symbols and the grammatical rules used to combine symbols into well-formed expressions. *Semantics* regards the meanings of the symbols and the methods for finding the meaning of expressions on the basis of the meaning of the parts of the expressions. *Reasoning* (often called proof theory in mathematical logic) is a method that produces additional expressions on the basis of an initial set of expressions.

On the basis of this definition, it is naturally deduced that language (syntax and semantics) is related to knowledge representation, while the proof theory is related to reasoning. So, there is a direct mapping between logic and KRR.

1.7 Artificial intelligence and bioinformatics

Bioinformatics is the discipline at the intersections of biology, computer science, mathematics, statistics, and physics that aims to analyze and interpret biological data [9]. It uses several data abstractions from computer science for modeling biological data (e.g., strings to represent nucleic acid or amino-acid sequences, 3D models to represent protein structures, even graphs to represent protein interactions networks), and in turn it applies specialized algorithms for extracting knowledge from such biological data (e.g., alignment for comparing protein sequences, structure prediction for predicting the secondary or tertiary structure of proteins, and network alignment for comparing protein interaction networks, to cite just a few).

In recent years, the availability of novel, high-throughput platforms for the investigation of cell machinery (e.g., next-generation sequencing, microarrays, and mass spectrometry), has given rise to the so-called omics world, characterized by the appearance of novel biological data (e.g., genomics, proteomics, metabolomics, interactomics, etc.) that are produced in huge quantities [10].

Omics data are gaining increasing interest in the scientific community because they are central for the understanding of the biological processes, as well as for the characterization of diseases, as happen for instance in the so-called P4 (predictive, preventive, personalized, and participatory) medicine [11].

High-throughput experimental platforms and clinical diagnostic tools, such as next-generation sequencing, microarrays, mass spectrometry, and medical imaging, are producing overwhelming volumes of molecular and clinical data, and the storage, integration, and analysis of such data is today the main bottleneck in the bioinformatics pipeline [12,13].

This big data trend in bioinformatics poses new challenges for both the efficient storage and integration of the data and for their efficient pre-processing and analysis [14]. Thus, managing omics and clinical data requires both support and space for data storage, as well as algorithms and software pipelines for data pre-processing, integration, and analysis. Moreover, as is already happening in several application fields, the service-oriented model enabled by the cloud is more and more spreading in bioinformatics.

On the other hand, the big data trend in bioinformatics is a great opportunity to improve the bioinformatics analysis pipeline by employing the power of AI. In fact, the availability of big omics data, together with the huge genome data made available by next-generation sequencing experiments, can feed the learning capabilities of AI algorithms.

Another technological trend that interests both bioinformatics and AI is the huge demand for computational power required by both disciplines. Parallel computing offers the computational power to meet the big data trend in bioinformatics and to support the high computational demand of AI and data mining algorithms [15]. Moreover, cloud computing is a key technology to mask the complexity of computing infrastructures, to reduce the cost of the data analysis task, and to change the overall model of biomedical and bioinformatics research towards a service-oriented model [16,17].

CHAPTER

Machine learning

2

CONTENTS

Abstract

Machine learning is a branch of AI concerning the development of algorithms and methods that enable computers to extract hidden and implicit knowledge from a vast amount of data. Several scientific areas can take advantage of machine learning, e.g., medical diagnosis, bioinformatics and cheminformatics, drugs development, medical imaging, and personalized medicine. Recently, the massive amount of produced biological data requiring analysis has exploded, spurring the development of many machine learning methods to deal with these massive amounts of available data. Research in life and omics sciences is performed manually by combining several statistical and mathematical tools. Due to the vast amount of omics data and the many possible combinations and integration of various omics data, conventional computational methods cannot work effectively and efficiently. Hence, machine learning can play a crucial role in providing efficient bioinformatics applications. This chapter introduces machine learning to create systems that can learn from data even to manage data never previously considered, e.g., genomics and proteomics data, providing the essential knowledge to understand the following chapters.

Artificial Intelligence in Bioinformatics. https://doi.org/10.1016/B978-0-12-822952-1.00011-5

Keywords

Classification, Clustering, Supervised learning, Unsupervised learning, Machine learning, Association learning, Reinforcement learning

2.1 Introduction

Learning is the process by which a system can improve its achievement through experience [18]. Roughly speaking, learning is the process of acquiring awareness from the surrounding environment (expertise) and converting experiences into knowledge models. From a computer point of view, the learning called Machine Learning (ML) is related to the algorithms and methods that enable computers to learn from input data [19]. Thus, for machines the surrounding environment is the input data set known as the training set, representing the *experience*. The output model is the *expertise* or, even better, the *knowledge*. The accuracy of such a pattern depends on the size of the input data set (experiences). More data to learn from (more experiences) make it possible to obtain more accurate and detailed knowledge (expertise). Thus, ML can help researchers in facing daily problems easily and more effectively.

Learning is, of course, a broad domain; consequently, it needs to group ML methods and algorithms to highlight the differences among the principal ML methods. ML can be organized into four main categories including **supervised** learning, **unsupervised** learning [20], association learning [21], and reinforcement learning [22], as depicted in Fig. 2.1.

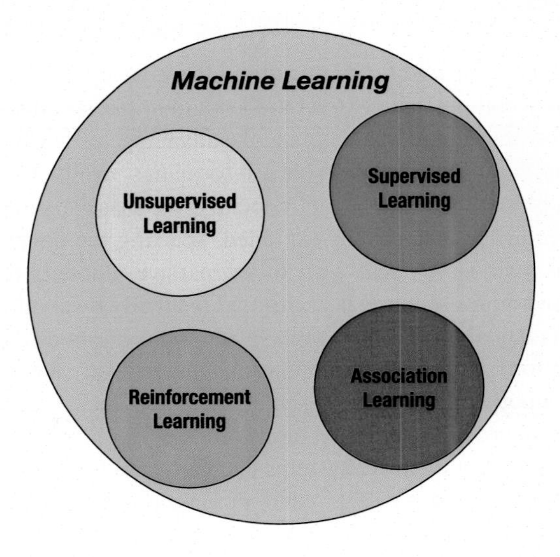

FIGURE 2.1

Graphical representation of machine learning categories.

Supervised Learning (SL) algorithms [23] can automate decision-making processes by generalizing from the input data, that is, experiences. In this context, the user must feed the algorithm with labeled input data known as the **training set**, including the target or class attribute. The training set contains examples to enable the computer to learn. The examples refer to the features or attributes provided to represent how to identify an item, an object, or an instance. The classification process finds a way to produce the desired knowledge model by exploiting the attributes. In particular, the classification process can be presented as a three-step process: *i)* production of a training model from a training dataset; *ii)* model accuracy estimation by using a test set; and *iii)* using the model to classify the unknown instances (e.g., instances whose class attributes are unknown). The classification methods can generate a model for an input that they have never seen before without any human assistance. For instance, early diagnosis is crucial in cancer research since it makes possible the classification of unknown patients into high- or low-risk categories for developing cancer [24]. Early cancer diagnosis and prognosis could improve the patient's subsequent management, contributing to the best chance for successful treatment. As a first step, researchers must extract the patients' historical data from a clinical database and then integrate clinical and genomic data to create the input dataset. As the next step, the dataset has to be split into two distinct sets: the training and test sets. The researchers also verify that only the training set contains the class attribute (*label*) necessary for the classification algorithm to learn. Remember that it is the class attribute that provides the information about the membership group, in this case, the low or high risk of each patient. The researchers can use the training set to feed the algorithm. As the last step to validate the model's accuracy, researchers use the test set that is different from the training set because it contains other instances, e.g., new patients with attribute values different from the training set. The algorithm should figure out a rule for labeling the new patients based on that training. As a result, the algorithm will then predict the cancer risk of the new patients.

In Unsupervised Learning (UL) [25], there are no known criteria on which to base the learning, rather something else is learned, e.g., groups, from the data structure without obtaining anything other than the data itself. In UL, often called clustering, there is no class attribute as in classification. Thus, UL's primary goal is to find a structural aspect of the data to analyze. Therefore, the problem is to investigate how points are organized in the input space. To effectively realize UL, a distance measure between instances of the dataset must be provided [26]. Common distance measures are the Euclidean [27] or Manhattan [28] distances. In unsupervised learning, there is no difference between training and test set. The UL algorithm processes input data in order to come up with some summary or compressed version of that data. The chief areas of applying UL algorithms include: data exploration to get intuition about the data by figuring out relevant groups between the data; recognizing similarities and structural differences between elements known as clustering; and dimensionality reduction (e.g., figuring out the more informative attributes) of a data set without losing too much information. Grouping a data set into disjoint sets of similar objects is a typical example of an UL algorithm.

For instance, a biologist can use clustering to create gene groups based on the gene-expression values. Hence, roughly speaking, we can define clustering as the task of grouping together similar objects, and dissimilar objects are distributed into various different groups.

2.2 Classification

Classification [29] attempts to yield the best possible knowledge model for a given problem. The methodology responsible for figuring out this knowledge model is known as classification. The classification must be able to generalize from a labeled input data set (the *training set*) from which to get experience (*knowledge*) and use it to classify unseen data (the *test set*). The classification goal is to gain expertise from labeled data in order to be able to assign a class label even for unknown data, through a mapping function $\varpi()$ (2.1).

$$\varpi(d) = d \rightarrow l, \forall d \in D, \text{ and } l \in \Gamma \tag{2.1}$$

Given an input data set $D = \{d_1, d_2, \dots, d_m\}$ and a class labels set $\Gamma = \{l_1, l_2, \dots, l_n\}$, the mapping function $\varpi()$ defined in Eq. (2.1), provides the best mapping of a generic item d_i on a class label l_j.

Thus, classification algorithms are suitable for problems where all the data are labeled, and it is necessary to predict the label for unlabeled new data. Classification can be grouped in two main categories, known as *classification* and *regression*. On the basis of labeled data, classification learns the rules that make possible marking new unseen data. In other words, classification algorithms can build programs that can act as classifiers. An easy way to represent a classifier is a *decision tree*. We talk of classification problems [30] when the output variable is a category, such as "diseases" or "healthy", "respond" or "not respond". Instead, when the output variable is a real value, such as "temperature" or "weight," we have a regression problem. Classification can be employed in many real-world scenarios. For instance, classification is employed in protein sequences to learn the functions of a new protein. In contrast, regression [31] aims to predict a real value based on one or more attributes. For example, regression can predict if the probability of a subject having high blood pressure is influenced by attributes such as lifestyle, education, or age.

Classification pursues the aim to yield universal models from the training data, which means models that can make accurate predictions on unseen data but with the same features as the training set used to train the model. Thus, classification models must be as general and accurate as possible, while trying to make them as simple as possible, avoiding *overfitting*. Overfitting happens when the amount of information used to define the model is too specific concerning the used training set. Thus, the resulting model will work well on this specific training set, but it will not generalize on new data.

2.2.1 Supervised machine learning algorithms

In this section will be introduced some of the most popular supervised machine learning algorithms.

- **Linear regression** is used to study the linear relationship between a dependent variable (e.g., diabetes) and one or more independent variables (age, weight). When there are multiple input variables, the procedure is called multiple linear regression. We can apply linear regression if the dependent variables are continuous.
- **Logistic regression** is employed to predict the probability of belonging to a group, e.g., *diabetic:(Yes/No)* and it is used if dependent variables are dichotomous, for example, *adverse drug reactions:(Yes/No)*. Logistic regression can be used to study the relationship among multiple dependent variables, providing as an outcome a measure of the association strength between the dependent variables and the rest of the variables.
- **Cox regression** is used to model survival data. It is applied when dependent variables indicate the survival time and the death time or current time, i.e., the time elapsed between the diagnosis to the event of interest. At the same time, the independent variables are continuous or categorical.
- **Decision trees** are supervised learning methodologies capable of making decisions based on a series of questions and answers. Decision trees can decide on a specific input through a series of questions, e.g., tests of the values assumed by the attributes, providing the best answers for the attributes' values. Hence, a decision tree takes as input an object defined by a set of attributes and returns a decision, i.e., the output value of that function. Each tree node contains a question, and each outgoing arch corresponds to a possible intermediate answer. The bottom of the decision tree encloses the leaf nodes of the final answers, that is, the predicted class. A path that starts from the root indicates a series of questions and answers leading to the leaves node is known as a decision rule.
- **Naive Bayesian** (NB) classifier is the simplest classification method since it is based on the hypothesis that the variables are themselves independent and the output variable is discrete. This means that a particular characteristic is unrelated to the presence of any other characteristic. The NB classifier can be grouped into three categories: multinomial, Bernoulli, and Gaussian. In multinomial NB, the data assume a multinomial distribution. Bernoulli NB is similar to multinomial except that the predictors are Boolean (True/False). Gaussian NB assumes that the continuous values follow a Gaussian distribution.
- **Neural networks** are algorithms that, unlike traditional algorithms, are not made up of a set of fixed instructions but are made up of mathematical models that allow networks to learn dynamically and without human supervision, providing even complex and unexpected outputs based on the provided input data. The mathematical neural network learning model requires on a training time that varies according to the complexity of the problem to solve. Neural network learning is based on the error backpropagation method [32]. The error backpropagation induces the neural network to retry several times, changing the connections' weights

and other parameters until the error is reduced or remains constant. In this manner the network can recognize unknown input data. Neural networks are distinguished according to the learning method used and can be grouped into three categories: Artificial Neural Networks (ANN), Convolutional Neural Networks (CNN), and Recurrent Neural Networks (RNN).

- **Support Vector Machines** (SVM) comprise one of the most often used SL algorithm able to gain knowledge in high-dimensional feature spaces. The SVM algorithm learns through the construction of a hyperplane that optimally divides a set of training data into classes. The hyperplane identification takes place through the following steps. Search for a hyperplane that separates the training data of one class from another. If there is more than one, SVM then chooses a hyperplane that improves the accuracy of the model, meaning the hyperplane with the largest distance to the nearest training data point of any class. If such a hyperplane does not exist, a nonlinear mapping function is used to transform the data point space into a new space with a higher number of dimensions.

2.2.2 Support vector machines

This section introduces the SVM methodology. The term SVM is due to the *support vectors'* choice identified in the training set and used to define the separation hyperplane or hyperplanes in the features space. The support vectors' characteristic is that they represent the data points with the greatest information content in the space of the features of the training set. The basic idea of the SVM is to directly solve the problem of determining the separation surfaces between classes (hyperplane) and that it is as far as possible from the points it separates. Generally, the resolution of such problems is based on the divide-and-conquer paradigm, intended to simplify the problem into easier-to-solve problems. Instead, an SVM algorithm does the opposite by adding complexity, in this specific case, adding a further dimension making the problem more manageable. The main SVM methodologies categories are *linear* and *nonlinear*.

Linear SVM can choose the best hyperplane that separates the two classes through the data points vector yielded from the training set. The chosen hyperplane must leave as many data points as possible from the same class in the same half-plane by maximizing the distance of the two classes' points from the hyperplane. Analyzing Fig. 2.2, it is worthy to note that there exist an infinite number of hyperplanes that can separate the existent data points. The points lying on the dashed line in Fig. 2.2 represent the separation vectors, also known as *support vectors*, the elimination of which would involve a change of the solution. The support vectors represent the training set's critical points because they are very close to the decision margin. The model produced by the classification depends solely on that subset of data points in the training set. Since it may happen that not all data points are separable from a hyperplane, it is necessary to relax the separation constraint so that the least number of data points can cross the class boundary.

Nonlinear SVM is based on the idea that training sets that are not separable in the *N-Dimensional* space of belonging may be separable in an higher-dimensional space.

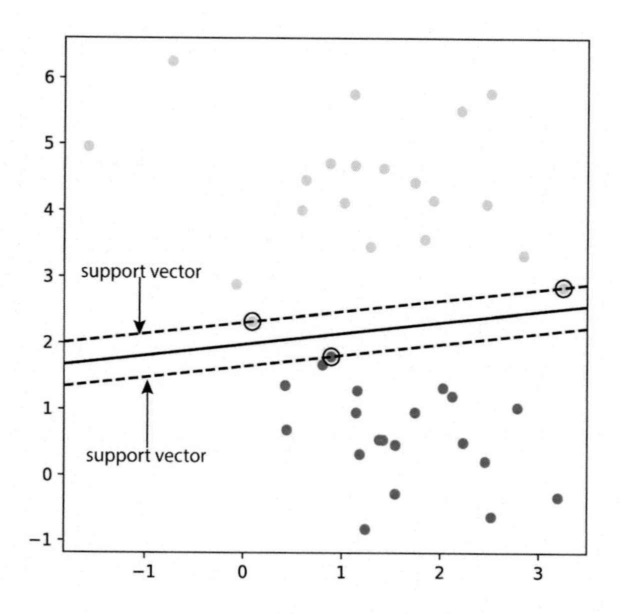

FIGURE 2.2

Representation of support vectors using the Python sklearn package [33,34]. The Python code is available at the following link: https://scikit-learn.org/stable/auto_examples/svm/plot_linearsvc_support_vectors.html#sphx-glr-auto-examples-svm-plot-linearsvc-support-vectors-py.

For this reason, SVM methodologies have been extended to specific classes with very complex surfaces. Nonlinear SVM to yield a new separable linear data points representation, employs a nonlinear mapping function to project data points in a new high-dimensional space. The nonlinear mapping function is also called the Kernel function $\kappa(\cdot)$. Nonlinear SVM can use various $\kappa(\cdot)$ functions to identify the hyperplane to separate the data points. The different kernel types are linear, polynomial of degree m, Radial Basis Function (RBF) similar to that used in Neural Networks, and customized functions defined for each specific problem. Fig. 2.3 shows an example of class separation using the $\kappa(\cdot)$ function.

2.3 Clustering

Clustering is one type of UL methods. The term unsupervised learning involves all the kinds of ways to extract knowledge from datasets that encompass unlabeled input data because the class attribute is not present. Hence, clustering is a class of learning algorithms based on observation instead of learning by example. Commonly, clustering concerns the separation of data points into several groups. Data points belonging to the same groups are similar, while data points in different groups are dissimilar.

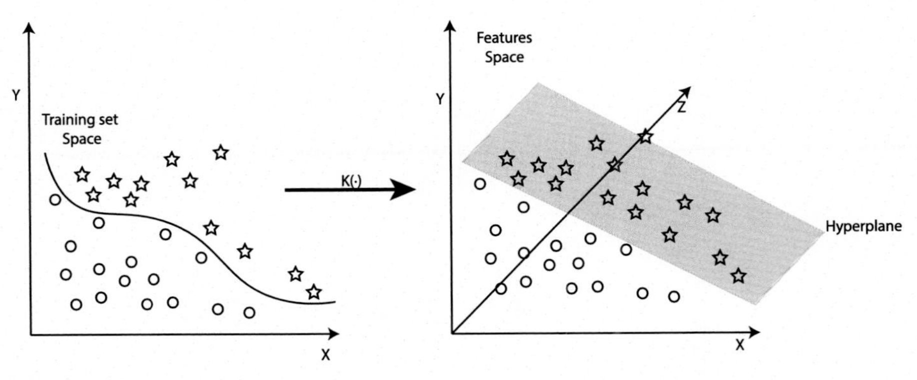

FIGURE 2.3

Example of SVM training set mapping through the $\kappa(\cdot)$ function in a new dimensional space.

The groups, e.g., clusters, are a collection of points obtained using similarity and dissimilarity functions.

Clustering methods can be arranged as follows:

- *Density methods* that determine the cluster assignment based on the concept of data density. Thus, a cluster is a dense zone of instances in the data space that is separated from the low density areas, i.e., with few instances. The Density-Based Spatial Clustering of Applications with Noise (DBSCAN) algorithm and the Ordering Points To Identify the Clustering Structure (OPTICS) belong to this category.
- *Hierarchical methods* that provide a hierarchical tree-type structure as a result. New groups are formed using the previously formed one. Hierarchical methods are divided into two categories:

  1. The agglomerative approach comprises AGglomerative NESting (AGNES). AGNES is an iterative bottom–up approach; at each step the nearest data points with respect to a distance measure are selected and merged. The process ends when all elements belong to a single cluster.
  2. The divisive approach includes DIvisive ANAlysis (DIANA). DIANA is a divisive top–down approach; it works in reverse order compared to AGNES. It starts with all instances in a single cluster. At each iteration, a cluster is split. The process ends when the individual instances stand alone in a cluster.

- *Partitioning methods* that split data points in k disjoint and not overlapping groups, each group forming a cluster. These methods require the user to define the number of clusters that are represented by the parameter k. These methods aim to maximize the clustering quality by employing similarity functions to optimize the objective. The *K-means* and *K-medoids* approaches belong to this clustering category.

- *Grid methods* that partition the data space into a finite number of cells to form a grid structure and form clusters from the grid structure's cells. In addition, the grid structure is employed to perform all the clustering operations.

The listed clustering approaches are based on the similarity and distance functions used to split an input itemset Υ into clusters. Given a set of items Υ, we can define a similarity or distance function over it. A similarity function f_S is a symmetric function where $f_S(y, y) = 1, \forall y \in \Upsilon$ and has a value in the range $[0, 1]$. A distance function f_D is a symmetrical function where $f_D(y, y) = 0, \forall y \in \Upsilon$. In addition to the similarity or distance functions, in some cases, it is necessary to define the number of expected clusters k. As a result, each approach will yield a partition of the initial itemset Υ into k disjoint clusters $\{C_1, C_2, \dots, C_k\}$. The produced clusters $\{C_1, C_2, \dots, C_k\}$ possess the following features: *i)* $\bigcup_{i=0}^{k} C_i = \Upsilon$, the union of all the clusters, will be equal to Υ, and *ii)* $\forall i \neq j \; C_i \cap C_j = \emptyset$, the intersection between each two different clusters will yield the null set.

2.3.1 K-means clustering algorithm

The K-means clustering algorithm can be defined as an optimization problem. The optimization problem consists of finding a partitioning (clustering) of minimal cost. Many standard objective functions require the number of clusters k as a parameter. The choice of the value of the parameter k is left to the user, who will provide the most suitable value for the given clustering problem.

In K-means, the data are distributed into k disjoint clusters $(C_1, \dots, C_k)$, that is, $C_i \cap C_j = \emptyset, \forall i \neq j$. Each cluster C_i is identified through a centroid Θ_i that indicates the center point of a cluster, as shown in Fig. 2.4.

The K-means objective function should measure and minimize the squared distance between each data point in Υ to the centroid Θ_i of its cluster. The K-means objective function has been widely studied within the scientific literature, so we can define the objective function over a data points dataset $f(\Upsilon)$ as stated in Eq. (2.2):

$$f(\Upsilon) = \sum_{i=1}^{k} \sum_{dp \in C_i} \| (dp, \Theta_i)^2 \| \tag{2.2}$$

Because in K-means the optimization criterion is to minimize the total squared error between the data points dp_i and their centroid Θ_i, the norm operator ($\| \cdots \|$) in Eq. (2.2) denotes the Euclidean distance. Euclidean distance measures the distances between two points in the space. Eq. (2.2) can also be rewritten as shown in Eq. (2.3):

$$f(\Upsilon) = \min_{\Theta_1, \dots \Theta_k} \sum_{i=1}^{k} \sum_{dp \in C_i} \| (dp - \Theta_i)^2 \| \tag{2.3}$$

The K-means is an iterative algorithm that alternates two main steps:

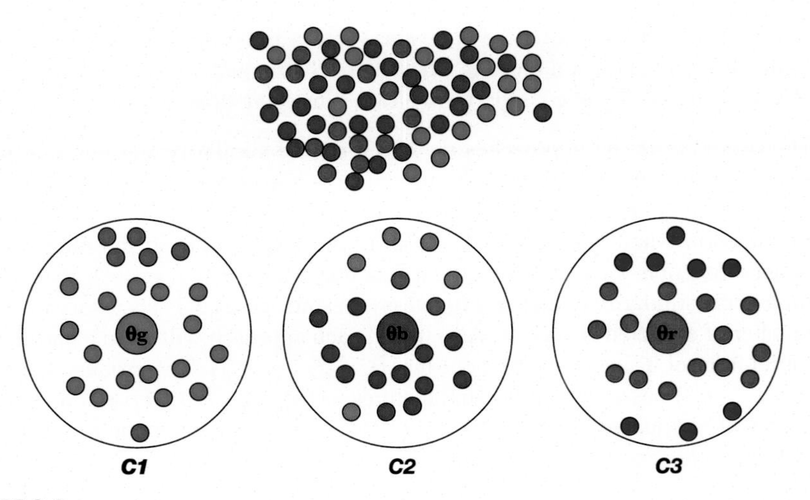

FIGURE 2.4

K-means data distribution, each cluster is identified through its own centroid (θ_g, θ_b, and θ_r). The three generated clusters are disjoint since $C1 \cap C2 = \emptyset$, $C1 \cap C3 = \emptyset$, and $C2 \cap C3 = \emptyset$.

1. For the given number of centroids k, assign each data point dp to its closest centroid using the *Euclidean* distance.
2. Update the centroids by computing the average of all the data points values assigned to it.

Algorithm 2.1 lists the main steps of the K-means procedure.

Algorithm 2.1: K-means pseudo code.

Require: the number of clusters k, a input dataset of items Υ
Ensure: k clusters.

1: **for** $i = 1; i \leq k; i++$ **do**
2: C[i] = selectDataPointsRandomly(Υ) {//randomly choose k data points from Υ as the initial centroids.}
3: **end for**
4: **repeat**
5: dp = getDataPoint(Υ);
6: assignDataPoint(dp);
7: updateClusterAvg();{//calculate the average value of the data point for each cluster}
8: **until no change** {//stop condition, e.g., the decrease in the objective function is too small.}
9: **end.**

The K-means algorithm converges because the objective function is monotonic. This means that after each iteration the objective function is non-increasing. The algorithm improves the centroids iteratively. When the centroids update does not yield any change in the objective function value, the algorithm has converged, and therefore it is not necessary to continue.

The sum of the squared error (SSE) is used to evaluate the correctness of the produced clusters. The SSE evaluates the clusters assignment's quality after the centroids converge or match the previous iteration's assignment. The SSE is defined as the sum of the squared Euclidean distances of each point to its closest centroid, using Eq. (2.3).

To elucidate how K-means algorithm works, we will analyze the following very simple gene expressions data set.

Table 2.1 A simple gene expression data set, showing the measured expression values of six genes.

	g_1	g_2	g_3	g_4	g_5	g_6
expression value	50.6	12.2	6.5	10.1	26.0	37.7

To group the gene expression data set in Table 2.1 with the K-means algorithm, we must provide the number of clusters K and randomly initialize the centroids' values. For instance, we can set $k = 2$, $\Theta_1 = 20$ for the first centroid and $\Theta_2 = 50$ for the second centroid. After the initialization phase, the K-means can start to compute the partition of the data points iteratively. In the first step, for each data point, the K-means algorithm computes the distance to Θ_1 and Θ_2, assigning the current data point to the closest centroid. For instance, 12.2, 6.5, 10.1, and 26.0 are closer to $\Theta_1 = 20$ than $\Theta_2 = 50$. Accordingly, they are placed together in the first group corresponding to the first centroid. The other data points, 50.6 and 30.7, are closer to $\Theta_2 = 50$ than $\Theta_1 = 20$. Thus, they are put into the second group. In this simple case, the closeness of a data point from the centroid can be calculated by subtraction. In the second iteration, the algorithm already converges; updating the clusters will yield the same results of the previous centroids. Fig. 2.5 summarizes the K-means main steps to build clusters.

2.4 Association learning

Association Learning (AL) [21] is the process of evaluating connections or associations among data according to their co-occurrence in a database or other data repositories. AL uses the Association Rule Mining (ARM) algorithms to discover unknown relationships between seemingly independently items stored in relational databases or other data repositories. Association rules help determine correlations between items and have applications in disease diagnosis, market basket analysis, and the retail industry. Association rules mining (ARM) was introduced by [35], and

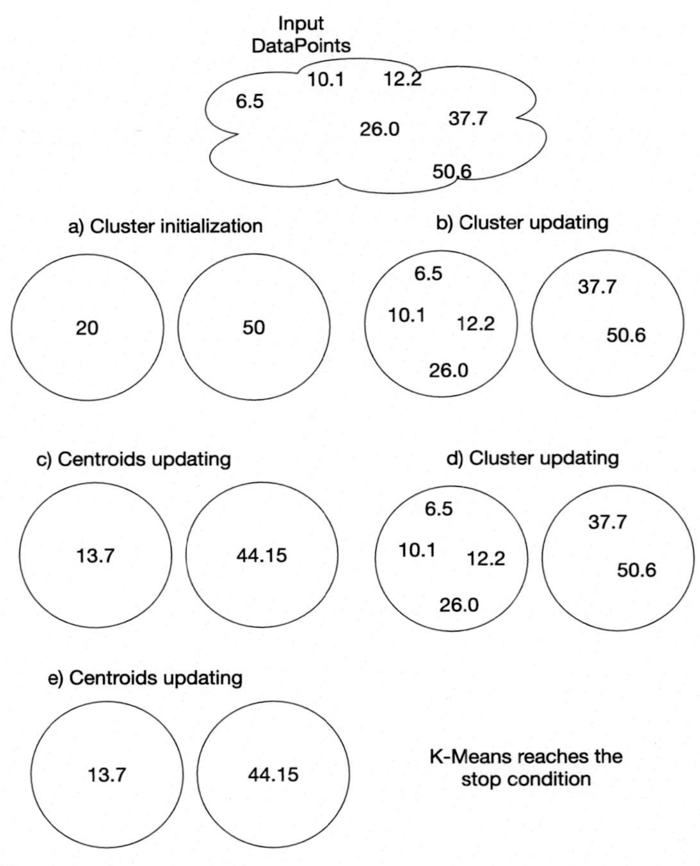

FIGURE 2.5

The schematic representation of K-means' main steps to build clusters.

it has attracted considerable interest because an association rule provides a concise statement of helpful information that end users easily understand.

Formally, the ARM problem may be stated as follows: Let $I = i_1, i_2, \ldots, i_n$ be a set of n distinct items. Let τ be the set of transactions, where each transaction t is a set of items such that $t \in I$. A set of items is called an *itemset*. The number of items in an itemset is known as the size of the itemset. An itemset of size k is called a k-itemset. An association rules mining problem can be decomposed into two parts:

1. to find all combinations of items in a set of transactions that occur with a specified minimum frequency threshold (called **Frequent Itemsets**).
2. to calculate rules that highlight the probability that one or more items are contained within frequent itemsets.

Association models calculate rules that express the relationships among items in frequent itemsets. For example, a rule belong to the frequent itemset composed by the following elements *A, B, C* might be stated as: (**IF (A $\wedge$ B) THEN C**) that can be read as: if A and B are included in a transaction, then C likely should also be included. In short, an association rule states the probability that an item or group of items implies the presence of another item. The **IF** part of an association rule is called **Antecedent**, while the **THEN** part is called **Consequent**. The intersection of the antecedent and the consequent is the empty set $IF \cap THEN = \emptyset$ since the two parts are disjoint and don't share any items. An example of an association rule looks like this: "`IF (bread and jam) THEN (peanuts butter)`." In market basket analysis, ARM methods can be used to figure out which products are most frequently sold together, providing information to improve the marketing strategy. In addition, ARM can be used in clinical applications to generate association rules from patient records. The obtained association rules can help medical workers to understand co-occurrences of symptoms associated with the potential to develop cancers, helping diagnose the disease at its earliest stage. The generation of meaningful rules can be regulated by using **Support** and **Confidence** measures. Support makes it possible to distinguish random rules from useful rules, while confidence is a trustworthiness measure associated with each discovered rule.

Formally, support and confidence can be stated as shown in Definitions 2.1 and 2.2.

Definition 2.1. Given two items X and Y, the *Support-Count* of X and Y, $S_c(X, Y)$ is:

$$S_c(X, Y) = \frac{S_c(X \cup Y)}{N}$$

Definition 2.2. Given two items X and Y. The *Confidence* of X and Y, $C(X, Y)$ is:

$$C(X, Y) = \frac{S_c(X \cup Y)}{S_c(X)}$$

The support-count is defined as the number of transactions containing both the itemsets X and Y, divided by the total number of transaction N. Instead, the confidence is defined as the number of transactions containing both the itemsets X and Y, divided by the number of transactions containing X.

A transaction τ is said to support an itemset $X \in I$ if it contains all items of X, i.e., $X \in T$, because an association rule is an implication of the form $X \Rightarrow Y$, where $X \in I$, $Y \in I$, and $X \cup Y = \emptyset$. The rule $X \Rightarrow Y$ holds in the transaction set τ with confidence C if $c\%$ of transactions in τ supporting X also support Y. The rule $X \Rightarrow Y$ has support s in the transaction set τ if $s\%$ of transactions in τ support $X \Rightarrow Y$.

2.4.1 Association learning algorithms

Apriori is the fundamental algorithm for finding frequent itemsets proposed by R. Agrawal and R. Srikant in [35,36] for mining association rules from databases.

Apriori is an iterative, incremental approach known as a level-wise search to identify frequent itemsets of incremental size. In the first step, the Apriori method scans the database to determine all the frequent items by computing for each item its support count. All the items that hold the constraint $S_c(i) \geq minSupp$ yield the $(k = 1)$-frequent itemsets. Next, the 1-itemsets are employed to uncover the 2-itemsets, and so on until it is impossible to extend k further, which means no more frequent k-itemsets can be found.

Apriori is a two-stage algorithm, *i)* candidate itemsets generation, and *ii)* rules creation. Algorithm 2.2 lists the main steps of Apriori.

Algorithm 2.2: Apriori pseudo code.

Require: the Transaction T counting the items I, and the minSupp value.
Ensure: L List of frequent itemsets of size k.
1: C_k {candidate itemset of size k}
2: L_k frequent itemsets list
3: $L_1 = scan(T)$ {Scans T to compute the support count for each candidate itemset}
4: **for** $k = 2$; $L_{k-1} \neq \emptyset$; $k++$ **do**
5: $C_{k+1} = gen_cand(L_k)$ {candidates generated from L_k}
6: **for** $t \in T$ **do**
7: C_t = getCand(t) {get the items of t that are candidates}
8: **for** $c \in C_t$ **do**
9: c.supp_count++
10: **end for**
11: **end for**
12: $L_k = c \in C_t | c.supp\_count \geq min_{supp}$ {Prunes all items for which $c.supp\_count < min_{supp}$}
13: **end for**
14: return L
15: **end.**

Having found the frequent itemsets from the transactions database T, it is straightforward to generate association rules. The rules that satisfy both minimum support and minimum confidence are called strong association rules.

The Apriori anti-monotonicity property can significantly reduce candidate sets' size in many situations, leading to good performance gain. Apriori anti-monotonicity property states that: "if any itemset with length k is frequent in the database, all its subsets must all be frequent." However, it may still be necessary to generate many candidate sets.

It may be necessary to repeatedly scan the whole database and check a large set of candidates by pattern matching. It is costly to review each transaction in the database to determine the support of the candidate itemsets.

The major bottleneck of the Apriori algorithm is the generation of frequent itemsets candidates because it requires multiple database scans. Is it possible to implement a method that mines the complete set of frequent itemsets to improve the candidate generation process? An interesting approach to accomplish this is called frequent pattern (FP) growth, or simply FP-growth, which embraces a divide-and-conquer strategy.

The advantage of FP-growth is that it can efficiently build an FP tree directly from the database, mining frequent patterns while avoiding multiple database scans.

FP-growth first compresses the database representing frequent items into an FP-tree, retaining the itemset association information. From the FP tree, the algorithm mines data in a bottom–up approach. The FP-growth algorithm encompasses two steps: *i)* build the FP-Tree by finding similar items called prefixes using two scans of the database; *ii)* mine the FP-tree recursively and generate frequent patterns from the intermediate FP-conditional pattern tree. The advantage of using the FP tree is related to the tremendous gain in reducing the data size to be searched in the mining process. This becomes possible because long frequent patterns share the same prefix, and the tree stores the shared prefix just once.

Algorithm 2.3 lists the FP-growth main steps.

Algorithm 2.3: FP-growth pseudo code.

Require: the FP-Tree FPT, $minSupp$
Ensure: L List of frequent itemsets
1: FP-Growth(Tree, t
2: **for all** $cT \in FPT.head$ **do**
3: $\beta = cT \cup minSupp$ {generate pattern with $support \geq minimumsupport$ of nodes in β}
4: $\beta = construct\beta()$ {construct β conditional base pattern}
5: $Tree_\beta = conditionalFPTree(\beta)$
6: **if** $Tree_\beta \neq \emptyset$ **then**
7: $FP - Growth(Tree, \beta)$
8: **end if**
9: **end for**
10: **end.**

2.4.1.1 Associative learning algorithms and software tools for handling genomics data

The existing algorithms, such as Apriori and FP-growth, are not very useful in mining association rules from genomic data to discover highly correlated genes that characterize an adverse drug response or to identify them. Apriori and FP-growth are suitable to handle numeric datasets like other machine learning algorithms. Hence, Apriori and FP-growth need to be customized to work with textual data to extract knowledge from genomics data. Pharmacogenomics investigations evaluate the amount of drugs that may cause lethal effects. A group of nucleotides known to be related to adverse drug reactions (ADR) has been determined in the past [37]. Consequently, investigating these polymorphisms may avoid incorrect drug intake preventing the consequent insurgence of adverse reactions since their presence/absence may favorite ADRs.

The drug metabolism enzymes and transporters (DMET) microarrays devised by Affymetrix have enabled the investigation of 1936 distinct nucleotides representing gene polymorphisms related to drug responses. DMET data produced in a single experiment are usually collected in a nxm matrix, where n is the number of probes and m the number of patients. Each entry i, j of such a matrix has the form A/T, in which the first letter represents the nucleotide of the first allele and the second letter represents the nucleotide of the second allele. In such a way, a genomic variant in two samples may be described in such a form: A/T for patient1 and T/T for patient2.

DMET microarray data are arranged as a high-dimensional matrix $M \times N$, such as Table 2.2.

Table 2.2 A simple DMET microarray data set.

	S_1	S_2	...	S_m
$Probe_1$	G/A	A/G	⋮	T/T
⋮	⋮	⋮	⋮	⋮
$Probe_2$	G/A	A/G	A/G	T/C

Since the association rules have been designed to process market basket data, they have to be adequately adapted to work with new kinds of data, such as those provided by DMET microarrays. This particular aspect is highlighted in [38], where the authors present DMET-Miner that is an extension of DMET-Analyzer [39] which is a software platform for the investigation of DMET data. DMET-Miner extends the FP-growth algorithm to mine association rules from DMET datasets. DMET-Miner defines optimized data structures with good performance results for rule extraction and massive pharmacogenomics datasets. In particular, the authors show how the proposed method outperforms off-the-shelf AR mining algorithms available in Weka and RapidMiner. However, the use of DMET-Miner on large omic datasets encounters some limitations. First, it needs a massive quantity of main memory that may limit the execution of DMET-Miner on personal computers. Second, a significantly longer response time may increase the time required to complete extensive pharmacogenomics studies. To overcome the limits of standard association rule mining algorithms implemented in DMET-Miner, the authors in [40] propose PARES (Parallel Association Rules Extractor from SNPs), a novel parallel algorithm for the efficient extraction of association rules from omics datasets. In addition, to further improve the performance, as well as the memory consumption of parallel AR mining algorithm, authors in [41] present a novel AR mining algorithm, called Balanced Parallel Association Rule Extractor from SNPs (BPARES), employing parallel computing and an efficient balancing strategy to improve execution time.

2.5 Reinforcement learning

Reinforcement learning [22,42] (RL) provides solutions to problems in which the system's output is obtained starting from a more-or-less complex sequence of actions or decisions, carried out in a dynamic environment with considerable uncertainty. In particular, RL can be exploited when the single action or the final result is not critical, but what matters is the proper sequence of steps to achieve the goal, also known as the policy or behavior. The RL should generate good policies/behaviors for solving the problems to be addressed, analyzing and exploring various autonomous actions, examining their validity, and learning from the consequences of previous actions. To this end, RL must use memory to archive what occurred in the past.

RL is an unsupervised machine learning method that generates policies/behaviors based on rewards and penalties. The rewarding procedures assign positive values in order to maximize cumulative rewards of the chosen actions to encourage their adoption. Contrarily, penalty functions give negative values to discourage the use of unwanted actions, enabling modifications or acquiring new behaviors and skills incrementally. Hence, both methods allow scheduling long-term activities to achieve good policy or behaviors, e.g., an optimal solution.

RL is an emergent and exciting topic in artificial intelligence (AI), but its adoption and application in the real world are limited. In particular, it is embraced in the following areas: gaming [43], robotics [44], recommendation system [45], and healthcare [46]. For example, consider a robot placed in an environment with limited perception that can perform actions like walking in one direction and employing a feedback mechanism. The robot interacts with the environment to perform the action and receive a response in a continuous cycle to improve behavior which is called trial-and-error experience. Thus, the robot does not need complete knowledge of the environment, and it only needs to interact with the environment and collect information.

In healthcare, RL could incorporate biologists and expert knowledge in analyzing complex diseases, such as cancer. Including biologists and expert knowledge could improve and speed up understanding the mechanism underlying cancer, learning possible actions from the expert through the reward function. In this way, the system can quantify the impact of expert decisions to provide effective therapies for cancer treatment.

CHAPTER

Artificial intelligence

3

The development of full artificial intelligence could spell the end of the human race. . . . It would take off on its own, and re-design itself at an ever increasing rate. Humans, who are limited by slow biological evolution, couldn't compete, and would be superseded.

Stephen Hawking told the BBC (December 2014)

CONTENTS

Abstract

We give an overview of the field of artificial intelligence. First, some various definitions of artificial intelligence are considered. Then, we discuss the evolution of artificial intelligence over time. Finally, we present the main steps of the introduction of artificial intelligence in bioinformatics.

Keywords

Artificial intelligence, Turing test, Artificial intelligence in medicine

3.1 A brief history of artificial intelligence

AI was introduced as an academic discipline in the 1950s, and it remained in a relatively minor field of research until the 2000s. AI can be classified into three major areas: analytical, human-inspired, and humanized AI, depending the kind of intelligence involved (cognitive, emotional, and social intelligence). Another classification that refers to its evolutionary stage has introduced three major classes: artificial narrow, general, and super intelligence.

Independent of the classification, there is a major question regarding all these approaches that also involves philosophical and ethical aspects. The question is that the behavior of an *artificial intelligent* machine should be indistinguishable from the human behavior.

Artificial Intelligence in Bioinformatics. https://doi.org/10.1016/B978-0-12-822952-1.00012-7

From an historical point of view, it is based on the distinction between dualistic and materialistic views of the mind. René Descartes in 1637 Discourse on the Method wrote:

"How many different automata or moving machines can be made by the industry of man ... For we can easily understand a machine's being constituted so that it can utter words, and even emit some responses to action on it of a corporeal kind, which brings about a change in its organs; for instance, if touched in a particular part it may ask what we wish to say to it; if in another part it may exclaim that it is being hurt, and so on. But it never happens that it arranges its speech in various ways, in order to reply appropriately to everything that may be said in its presence, as even the lowest type of man can do."

In summary, Descartes noted that automata are capable of responding to external stimuli but they cannot respond the way that any human can.

According to dualism, the mind is nonphysical, i.e., it has nonphysical properties and therefore a pure physical approach cannot explain the mind at all. In this scenario it is impossible to reproduce the behavior of the mind. Conversely, materialism argues that the mind can be explained physically, and therefore it is theoretically possible to reproduce the mind.

Starting from these considerations, in the middle of the 20th century researchers in the United Kingdom explored for many years the possibility to build *machine intelligence* and then founded the **artificial intelligence** field in 1956 [47]. This field was started by the members of the Ratio Club, a group of researchers including Alan Turing. In particular, the pioneering work of Alan Turing during WWII constituted the foundation stone of the concept of computer intelligence. In particular, Alan Turing investigated *the question of whether or not it is possible for machinery to show intelligent behavior*. To resolve the problem, Turing proposed a first experiment involving three subjects: A, B, and C. Then, in his paper *Computing Machinery and Intelligence*, Alan Turing in reasoning about the question 'Can machines think?' proposed to reformulate the proposition as: *Can machines do what we (as thinking entities) can do?*. Then, he proposed a test known as the "imitation game". In this test a man and a woman go into separate rooms, and guests try to tell them apart by writing a series of questions and reading the typewritten answers sent back. Both the man and the woman have the goal to convince the players that they are the other. Turing modified the game by exchanging one of the two players (man or woman) with a machine as depicted in Fig. 3.1. This enabled Turing to formulate the following question: "Will the interrogator decide wrongly as often when the game is played like this as he does when the game is played between a man and a woman?" [48].

After this seminal work, the term artificial intelligence was introduced officially in 1956 in a conference at Dartmouth College which included for the first time a session titled artificial intelligence. The LISP (List Processing Language) is a functional programming language that was developed to build programs for AI.

After this first period, AI had a dark period until 1970. Between 1970 and 1975, AI gained the attention of researchers, and it achieved some success related to a few

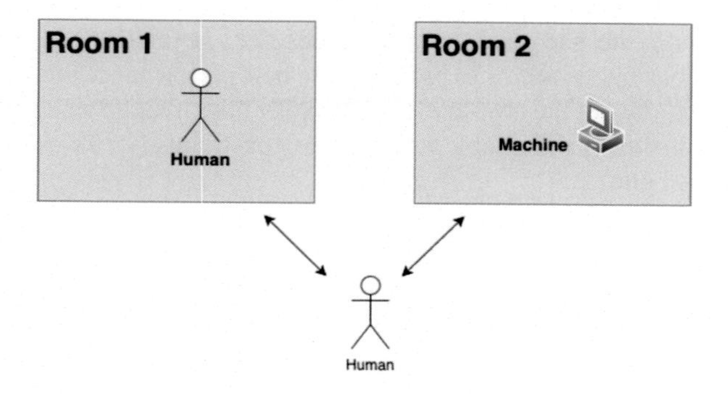

FIGURE 3.1

In the imitation game there are three subjects: a human in a room, a machine in a second room, and a human acting as the interrogator. The interrogator has to discover which is the human being and which is the machine, while the goal of the machine is to be indistinguishable from a human being.

applications (see the next section). This reinforced the idea that AI could expand into other branches of science, such as psychology, biology, and medicine.

3.2 Artificial intelligence and bioinformatics

The application of Artificial Intelligence (AI) in bioinformatics, computational biology, and medical informatics is continuously growing. There is expanding interest motivated by the consideration that many of the problems in bioinformatics need AI because of given the computational complexity of existing methods or the impossibility to discover hidden links and relationships within biological data.

For instance, protein structure prediction, i.e., the prediction of the three-dimensional structure of the proteins given a sequence of amino acids [49,50], is known to be a problem requiring long computational time. The AlphaFold[1] project represents a successful story of the use of AI to deal with this problem.

The possible applications of AI span a broad spectrum, from reverse engineering of genes and metabolic networks to experimental data overwhelming the existing limitation of standard techniques for analyzing gene expression and proteomic data, such as data clustering [51,52]. Similarly, many existing methods may benefit from the integration of further biological knowledge such as human-in-the loop AI or similar.

[1] https://alphafold.ebi.ac.uk/.

AI (in particular, unemotional AI) is an area of computer science which originated 1950s, and it initially specialized in dealing with complex problems that were intractable through the use of classical methods (i.e., heuristics or algorithmicsapproaches). AI, conversely, seems to produces the best performance when dealing with problems that require not the absolutely best answer but rather an answer that is better than the previous one or acceptable under some constraints (i.e., weaker constraints w.r.t. the previous one).

3.3 Artificial intelligence in medicine: a short tale

The application of AI in medicine has grown very rapidly over the previous 20 years, with a focus on creating conditions for making personalized medicine possible. One of the main goals of these applications is to build predictive models that improve diagnoses of diseases, while tailoring therapies to the patients and also fostering preventive medicine, i.e., to avoid the insurgence of disease. In parallel, AI is used to enhance diagnostic accuracy with technological platforms (e.g., enhancing the diagnostic power of medical images) and to improve clinical operation (e.g., suggesting therapeutic or hospital workflows). The history of AI in medicine is characterized by the introduction of seminal applications over successive time periods.

As summarized in [53], the early period for AI in medicine is from the 1950s to 1970s. In this period AI focused on the realization of *machines* able to make inferences in the place of the humans. A seminal example of these applications is the robot arm called Unimate (developed in 1961 by Unimate Inc.; Unimation USA) [54] that was able to perform step-by-step commands. Furthermore, the Stanford Research Institute created Shakey in 1966, a robot able to interpret and execute simple one-step commands [55]. Furthermore, the software called Eliza was introduced to communicate using intelligent techniques simulating human conversation [56]. In parallel some independent projects introduced clinical informatics and bioinformatics databases (i.e., the PubMed search engine) and medical record systems that were used as main data sources for future projects.

After the pioneering decade, AI in medicine experienced a substantial lull in the 1970s to the 2000s. This period was characterized by limited funding and research. Despite this time period, two major advances should be reported since collaboration among experts continued even with the limited funds available. One of the first remarkable prototypes was the consultation program for glaucoma developed on the framework of the CASNET model of Rutgers University. This model was based on three main programs: a model building software, a consultation module, and an underlying database of data and models. The program aimed to support physicians in patient management [57].

Conversely, some seminal but fundamental advancements occurred from 2000 to 2020. In 2007 IBM developed Watson which was an open-domain question answering system based on DeepQA, a technology that leveraged natural language processing and ad-hoc search strategies to analyze data to generate probable answers to the given questions [58]. Watson was used to support clinical decision making and to predict proteins that were related to amyotrophic lateral sclerosis (ALS) [59].

Moreover, in 2000 Deep Learning (DL) experienced unique interest in the community because the introduction of computers with high-computing power made DL feasible even on large medical datasets. DL in medicine was introduced for the processing of medical images, where convolutional neural networks (CNNs) were largely used. For instance, CardioAI was able to analyze cardiac magnetic resonance images supporting diagnosis of lesions or the analysis of functionality. CardioAI was also successfully used for X-ray images for the chest, liver, and lung and also CT images of the head [60]. AI was also used in brain imaging [61,62].

CHAPTER

Data science

4

CONTENTS

Abstract

In the last few decades we have witnessed an acceleration of the process that aims to essentially render every aspect of reality in a form suitable for computational analysis, i.e., data. Thanks to the development of technologies capable of carrying out a massive number of tests and immediately acquiring a massive amount of data, even the biological sciences have undergone a profound transition from mainly driven-by-hypotheses disciplines to data-driven sciences. Data science is the field of study that encompasses the scientific principles, methodologies, and techniques suitable for the collection and the computational analysis of data, in order to draw from these new insights, i.e., novel information that is understandable, not obvious and not known a-priori, that can be useful in decision-making processes. We will introduce data science, briefly discussing the stages of a general data science process. Finally, various platforms and programming languages for data science will be presented.

Artificial Intelligence in Bioinformatics. https://doi.org/10.1016/B978-0-12-822952-1.00013-9

Keywords

Data science, Machine learning, Bias-variance decomposition

4.1 Introduction

Thanks to the development of new technologies, we are now capable of acquiring, storing, and annotating massive amounts of data.

Data can be defined as a set of measurements, also called features, relating to the same entity or the same event, collected to describe it. Data, therefore, express an abstraction of reality represented by measures, categories, symbols, images, signals, etc. In recent years, more than ever, data has attained a very high value. This is because their analysis is seen as the key to providing "actionable insights" into a plethora of real-world problems that we are interested in addressing.

This is why in the last few decades we have witnessed an acceleration of the process that aims to essentially render every aspect of reality in a form suitable for computational analysis, i.e., data. This process, different in scope from digitization, has been called *datafication* [63,64].

In healthcare, the process of datafication has spanned several sectors. These include public health infrastructures; biobanks, precision medicine; the data produced by implantable biosensors, by tele-healthcare devices, IoT devices and wearable devices, up to social media platforms and self-care practices [65].

Data science is the field of study that encompasses the scientific principles, methodologies, and techniques suitable for the collection and the computational analysis of data, in order to draw from these new insights, i.e., novel information that is understandable, not obvious and not known a-priori, that can be useful in decision-making processes.

The actionable insights, also called patterns or more generally new knowledge, can be of various types. A first division is commonly made between descriptive and predictive patterns. Among the descriptive patterns, one may be interested in the recognition of subsets in data that exhibit some kind of similarity, defined in terms of distances between the data, and are associated with clustering tasks. Alternatively, one may be interested in determining relationships or dependencies between features, which can be defined through the "if X, then Y" rule, and this task is known as associative rule mining.

If, on the other hand, the goal is to infer a measurement or a category on the basis of already known examples, it is a case of predictive or discriminative patterns. Predictive patterns are generally associated with classification tasks, although we tend to distinguish cases where the feature of interest, or feature target, is a class from cases where the target feature is a numerical measurement (regression).

Another popular definition, presents data science as that field of study at the intersection of statistics, machine learning, data mining, and computer science [66,67].

The difficulty in defining data science to differentiate it from other fields of study lies in the fact that many of the key elements of this discipline have been developed in related fields, i.e., statistics, data mining, and machine learning. On the other hand, it is necessary to point out the differences between data science and these fields in order to better understand it.

Historically, the term data science was born around the end of the 1990s to indicate the need for an evolution of statistics that would benefit from computer science [68]. In particular, this "renewed statistics" had to exploit the growing accessibility of data extracted from massive relational databases [69], together with the availability of algorithms and computational models for data analysis, such as neural networks [70].

The debate on whether or not to identify data science with statistics has sparked scientific interest for at least a decade [71]. In the last 20 years, thanks mainly to the ever-increasing amount of data (big data) generated especially online and to the rising interest in the possibility of analyzing them, the field of data science has rapidly grown, distancing itself from the concept of the "new statistics". Currently, data science has acquired a multidisciplinary connotation, strongly linked to the application domain, which makes it the natural extension of data mining.

The terms data mining and data science are often used interchangeably. Data mining, also known as knowledge discovery in databases, emerged between the end of the 1980s and the beginning of the 1990s. In a famous article by Fayadd [72], it is defined as:

"The non-trivial process of identifying patterns in data that are valid, original, potentially useful and understandable."

This definition is very similar to the one used at the beginning of this chapter to introduce data science, although data science is a field of study, while data mining refers to the process of identifying patterns in data.

As anticipated, we can consider data mining as a subfield of data science as the first deals with analyzing structured data—in general for commercial purposes—while the latter aims to appropriately collect a set of data (structured and unstructured) and extract from these those patterns that provide useful information about a problem. Data science typically addresses problems on a larger scale than data mining and for research purposes. One of the major contributions of data mining that is used in data science, albeit with appropriate modifications, is the Cross Industry Standard Process for Data Mining (CRISP-DM). The CRISP-DM is a general semistructured process model, which is software and problem-agnostic and encompasses six steps that naturally describe every data mining life cycle.

One core step of the CRISP-DM model, also present in the data science life cycle, is the modeling step that essentially automatically builds models from data by drawing on machine learning approaches.

Table 4.1 Main differences between data science and machine learning.

	Data Science	Machine Learning
Key element	Data	Algorithm
Human factor	Central	Algorithm design
Domain-dependency	High	Low
Goal	To extract new insights from data	To train an automated system to perform complex tasks based on the collected data and improve from experience

Also in this case, the terms data science and machine learning are often used interchangeably, however there are substantial differences between the two fields, as summarized in Table 4.1.

First, if the primary goal of data science is to extract new knowledge from data, machine learning aims to enable computers to learn from data without being explicitly programmed to do so[1] [74,75]. Therefore, if the former places data at the center of all activities, the focus of the latter is the algorithm.

In data science, the human factor plays a central role, not only in terms of knowledge of the domain, and in the interpretation of the results of the analysis, but also in each of the steps of the process. In fact, based on the result of one of the stages of a data science process, it is the data scientist who decides whether to move towards the next step or return to one of the previous steps to try to improve the results of the current step. On the contrary, in machine learning it is assumed that the learning process is automatic, and the human factor only comes into play in the design or optimization phase of the algorithm.

Typically, the evaluation of the plausibility of a machine learning model is based on consolidated metrics and on a gold standard dataset. In addition to these practices, a good data scientist should also evaluate whether the results obtained in a given process, in addition to being plausible, find theoretical confirmation. Therefore, the data scientist should have enough knowledge of the application domain combined, possibly, with the possibility of interacting with the experts of the domain in order to fully understand the problem and identify an appropriate solution.

Another key point for data science is therefore the interpretability of the patterns extracted from the process. However, many machine learning models, and in particular deep learning models, are often considered as "black boxes". For this reason, the scientific community's need to make machine learning models more interpretable has grown in recent years. More details on explainability will be discussed in Chapter 7.

In the remainder of this chapter we will discuss the stages of a data science process, focusing on classification tasks. Finally, various platforms and programming languages for data science will be introduced.

[1] This definition is often attributed to Arthur Samuel [73].

Table 4.2 Confusion matrix.

Category	Data Type	Description	Example
Numerical/ quantitative	Interval	Numerical scale with significant intervals	Temperature in Celsius scale
	Ratio	Interval with true-zero	Weight
	Discrete	Integer	Number of pregnancies
Categorical/ qualitative	Ordinal	Discrete variable with ordering relation	low, medium, high
	Nominal	Discrete variable without ordering relation	Healthy, pathological

4.2 A quick primer on data

As the name suggests, data science basically revolves around data. We can define datum as a variable, feature or attribute, i.e., any observation collected to measure or determine a given feature or event. The methods and techniques of measurement can be extremely sophisticated, as for example the ones produced by IoT systems, or texts extracted from medical reports; or they can be very simple as for instance the presence or absence of a symptom in a patient; and they can be explicitly measured or derived. The average number of daily COVID-19 infections in March 2020 is an example of a derived measure.

Generally, however, it takes a set of measurable and distinct characteristics to describe an entity or phenomenon.

For example, to describe a patient with diabetes, various measurements made on the patient can be considered, such as the blood glucose levels, the body mass index, and the number of pregnancies she has had in her life, as well as a series of specific exams results.

As can be seen from the example, features can be of a different nature and their nature affects the methods to be used to understand and manipulate data. It makes sense to say that a person has a BMI of 25.5, while it generally makes no sense to say that she has had 2.5 pregnancies.

The taxonomy of the most frequent feature's types is summarized in Table 4.2. A first distinction is made between quantitative and qualitative feature. The former represents quantifiable measures in terms of discrete or real values. In turn, real values, which make sense in an arbitrary interval, can be measured on a scale that has a true-zero origin (ratio scale) or not (interval scale). Typical examples of ratio scale measurements are pressure or temperature on the Kelvin scale, while an example of interval scale measurement is temperature in degrees Celsius.

Qualitative features take values in a finite set of labels, which represent different properties of that entity. If labels can be ordered in an increasing or decreasing fashion, the feature type is called ordinal, otherwise nominal. A feature that expresses the possible stages of a disease is an ordinal feature, while a feature that expresses the color of the wine is nominal. Among the nominal features, binary attributes (e.g., healthy/pathological subject) are of particular importance.

A common mistake is to confuse the feature type with its encoding. A binary nominal attribute can be coded numerically, for example, by setting healthy subjects = 0 and pathological subjects = 1. However, it must be kept in mind that codes cannot be treated as real numbers since, for example, if there exists an order between code 0 and code 1, the order does not hold for the two nominal classes.

If our goal is to assess whether it is possible to predict whether or not a patient suffers from diabetes based on the features already listed, we cannot extract this knowledge from a single patient, but it is necessary to know the characteristics of a sample of patients that may include both cases of diabetic and non-diabetic patients. In general, a collection of examples relating to a particular problem is called a dataset.

In its simplest form a dataset is structured, i.e., it can be represented through a matrix of N rows and D columns, in which each row represents a different patient (more generally a different instance of the problem), while the columns contain the set of defined features and represent the initial dimensionality of the problem.

Most of the data have an unstructured nature—think of a collection of textual datasets, or images, or biosignals. However, also to extract models from unstructured data, it is necessary to give as input to machine learning algorithms a matrix representation of the dataset. Structured features are extracted starting from unstructured data by using suitable techniques that are specific for the data source and, in some cases, also for the specific application domain.

Therefore, in the rest of the chapter the main focus will be on structured data because the same steps and consideration will ultimately apply to unstructured data as well.

To date, some among the most popular computational data formats are:

- CSV (comma separated values). Plain text files in which each line represents a single instance and features are separated by commas.
- TSV (tab separated values). Plain text files in which each line represents a single instance and features are separated by a tab.
- XML (eXtensible Markup Language). A metalanguage for the definition of markup languages, i.e., a language based on a syntactic mechanism that allows one to specify the meaning of the elements contained in a document or in a text. Unlike HTML, the creation of custom tags is allowed.
- JSON (JavaScript Object Notation). It is a format for the exchange of data objects between client–server applications. It is improperly compared with XML which, however, is a markup language, as it is composed of a list containing attribute-value pairs.
- HDF5 (Hierarchical Data Format version 5). It is a hierarchical file format designed to store and organize large amounts of heterogeneous data. HDF5 file format includes two main types of objects: datasets, which are homogeneous multidimensional arrays, and groups, which can contain datasets and other groups. Both groups and datasets have metadata associated with them. One of the reasons for the popularity of this format is the availability of libraries that support reading and writing it.

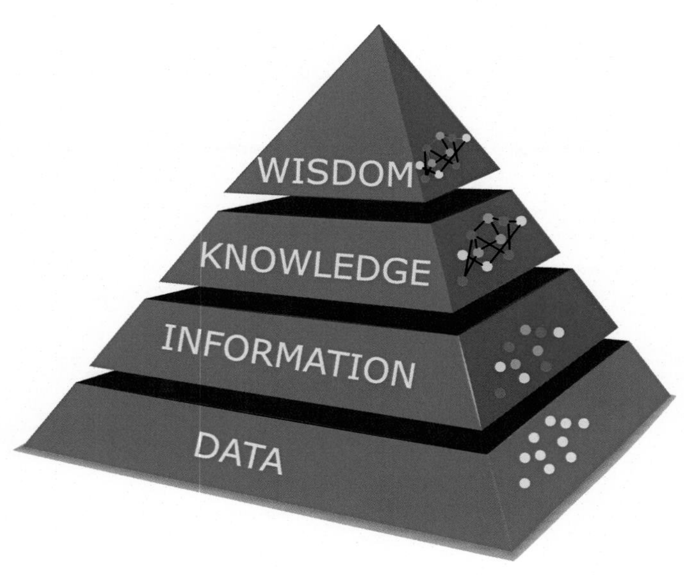

FIGURE 4.1

The knowledge hierarchy: raw data are abstract representations of the world; information is preprocessed data to make it easier to structure, visualize, and analyze; knowledge is information with meaning that makes it useful; wisdom is applied knowledge [76,77].

4.3 The data science process

As anticipated in the introduction, data science can be considered a generalization of data mining as the knowledge discovery process designed for structured data is extended to semi-structured and unstructured data, and data science is broader in scope.

In fact, many key concepts of data mining are present in data science. Of particular importance is the knowledge hierarchy, also called the DIKW (Data, Information, Knowledge Wisdom), pyramid (see Fig. 4.1), which schematizes the process that from data leads to wisdom passing through information and knowledge, and the CRISP-DM model, which outlines a knowledge discovery process regardless of the software, application domain, problem, or particular algorithm chosen for modeling. Fig. 4.2 depicts the data science life cycle based on the CRISP-DM model.

4.3.1 Problem and data understanding

One of the central themes in any data science problem, as indeed in any field of knowledge, is framing the problem of interest and finding the right data for a (data-driven) solution. For this reason the first two steps of the CRISP-DM model will be discussed jointly.

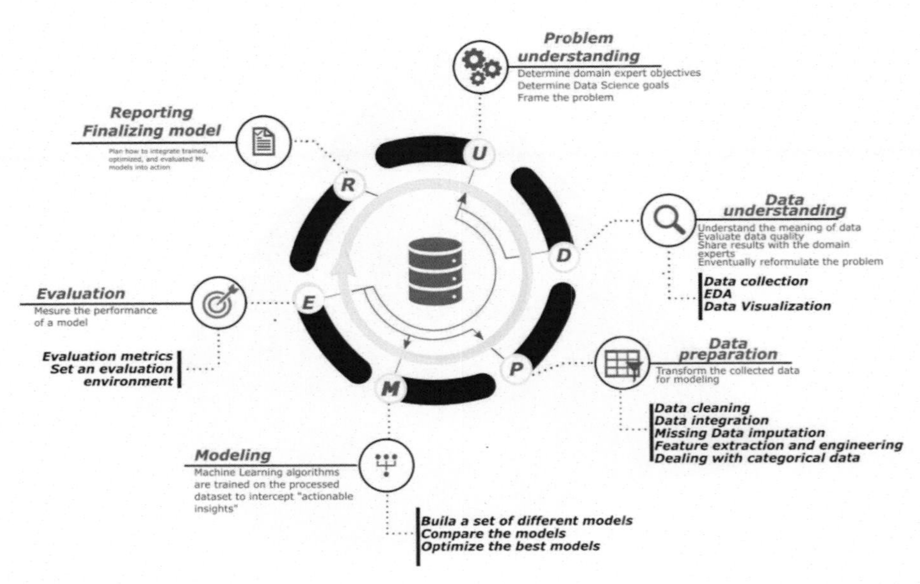

FIGURE 4.2

The data science life cycle.

To frame the problem, it is necessary to discuss with the domain experts and understand the research questions and analysis objectives of interest to them.

Another good practice is learning about the application domain, including through state-of-the-art research. Another fundamental activity to understand the problem is to determine if the domain experts made available the appropriate data for a data-driven solution to the problem or, alternatively, to evaluate the availability of useful data sources.

There are currently several platforms that share quality datasets, e.g., Kaggle,[2] government resources, and academic datasets, available for example on UCI.[3]

An almost inexhaustible source of data consists of web pages and social media platforms containing texts that can be collected through web scraping, spidering, and crawling, facilitated by libraries available for various programming languages. Clearly, data collection must always take into consideration the limitations of the particular web service, together with ethical issues that include anonymization of data and many others.

Every data scientist has heard the phrase "garbage-in, garbage-out" at least once. It constitutes a reminder that poor quality data results in poor quality models. For this reason, once the data has been collected, it is necessary to explore the data collection in order to:

[2] https://www.kaggle.com/datasets.

[3] https://archive.ics.uci.edu/ml/datasets.php.

- better understand the characteristics of the data you are dealing with, in terms of dimensionality and number of the dataset, and understand the meaning of the various features present and their respective types;
- evaluate the quality of the data collected, detecting any errors that occurred for example during the acquisition process;
- share the initial results of the exploration of the data with the domain experts and reformulate the problem if advisable.

This step is known as Exploratory Data Analysis (EDA) and exploits descriptive statistical tools, such as frequency measures, central trend measures (mean, median, mode), dispersion measures (variance, interquartile range, etc.), and measures related to the shape of data distribution, such as skewness[4] and kurtosis.[5]

Another tool lent by statistics is the correlation analysis between features.

However, a statistical summary may not identify some issues in the data, or it may be difficult to read for domain experts. Therefore visualization plays a central role in EDA. The visualization can be both static, leveraging high-quality graphs and plots, or interactive through dashboards.

The visualization of the data, although somewhat limited by the type of features present in the dataset, in a certain sense gives an artistic connotation to the EDA process: it is very likely that different data scientists will produce different visualizations of the dataset, focusing on different aspects yet ones complementary to understand the data.

An excellent discussion related to good data visualization practices can be found in [66].

4.3.2 Data processing and feature engineering

The objective of data preparation is to transform the collected data, which could be structured, semi-structured, or unstructured, in a form suitable for the subsequent modeling step.

In various surveys compiled by data scientists, it emerges that collecting, exploring, and preparing data occupy a massive proportion of the data analysis, ranging from 60 to 80%.[6] Especially in the data preparation phase, in fact, various potential issues need to be resolved before data modeling. Clearly, it is possible to identify generic classes of problems related to data preparation, but the specific technical resolution strongly depends on the data source being analyzed.

The most common classes of problems include:

- Data cleaning;
- Data integration;

---

4 Skewness measures the lack of symmetry in a distribution.

5 Kurtosis measures whether the data distribution is heavy or light-tailed w.r.t. a normal distribution.

6 https://www.anaconda.com/state-of-data-science-2020.

- Missing data imputation;
- Feature extraction and engineering;
- Dealing with categorical data.

Data cleaning deals with the removal of artifacts in the data, such as inconsistent notations, numerical values falling outside the range of admissible values (outliers), spelling errors in textual data, noise in biosignals, etc. Data integration focuses on making data from different sources comparable, for example by unifying units of measures, unifying notations, etc.

Although machine learning algorithms do not permit missing values as inputs, these are generally present in real-world datasets. Moreover, contrary to common belief in the big data era, the majority of a dataset, for instance, data collected from clinical trials, do not contain enough data to allow the removal of records with at least one missing value. For this reason, a common practice is to replace missing values with estimated values. Estimated values can be imputed by statistical measurements of central tendency, by exploiting data similarities, or through predicted values from regression models.

Feature engineering refers to the process of using domain knowledge to extract the most relevant features that are suitable for data modeling from raw, usually unstructured, data.

4.3.3 Modeling

In this phase, machine learning algorithms are trained on the processed dataset in order to build a model that can intercept the patterns hidden in the data and exploit them, for example, to make predictions. In Chapter 2 an overview of the most popular machine learning algorithms was presented. It would make sense to ask if there is an "oracle" algorithm among them that on average achieves higher performance than others on all possible problems.

The answer to the question has been formulated in a theorem, known as "No free lunch theorem" by Wolpert [78,79], which demonstrates that in optimization theory "the average performance of any pair of algorithms across all possible problems is identical". Repairing Wolpert, for any two algorithms A and B, if A outperforms B on a class of problems, there will be a class of problems in which B outperforms A.

Leaving aside the errors due to the quality and generality of the data, what are the errors that a machine learning model makes? They are essentially of two types: errors related to high learning bias and high variance. The first is a bias due to intrinsic assumptions in the algorithm, e.g., a linear model produces a high bias in modeling non-linearly distributed data, while the variance indicates how sensitive the model is to the fluctuation of training data. High bias produces underfitting models: The complexity of the model is too low to model the data, while high variance produces overfitting models: The model is too complex and fits "too well" to the data on which it was trained, that fail to generalize over unseen data.

The goal is to determine a model that achieves a bias–variance trade-off. In practice, this translates into:

- firstly build different models starting from different algorithms,
- compare the various models to understand which model outperforms the others,
- optimize the models with the best results.

To make sure that the model comparison provides a realistic and general estimate of the performance of the model itself, two key ingredients are needed:

- a collection of metrics that estimate the goodness of the model;
- a suitable evaluation environment.

4.3.4 Evaluation

4.3.4.1 Evaluation metrics for binary classification

Measuring the performance of a classifier means quantifying how many and which instances are predicted correctly and how many are predicted incorrectly.

Let us return to the dataset of diabetic and non-diabetic patients, indicate with Positive the presence of diabetes class and with Negative the absence of diabetes class, and suppose that we have built a prediction model on the data. Given a record as input to the model, four cases can occur: The record belongs to the Positive class and is recognized as Positive or misclassified as Negative; the record belongs to the Negative class and is correctly predicted as Negative or misclassified as Positive. Indicating with True positive (TP), True negative (TN), False positive (FP), and False negative (FN) respectively the number of positive/negative instances predicted correctly and the number of positive/negative instances predicted incorrectly, it is possible to represent the situation with the matrix in Table 4.3, called a confusion matrix.

Starting from the confusion matrix, different measures can be defined. Accuracy indicates the portion of the dataset that has been correctly classified:

$$Accuracy = \frac{TP + TN}{TP + TN + FP + FN}$$

Precision is the portion of correctly classified positive instances with respect to the total of positive predictions made by the model:

$$Precision = \frac{TP}{TP + FP,}$$

while Recall is the portion of correctly classified positive instances with respect to total instances actually positive:

$$Recall = \frac{TP}{TP + FN}$$

Table 4.3 Standard features type.

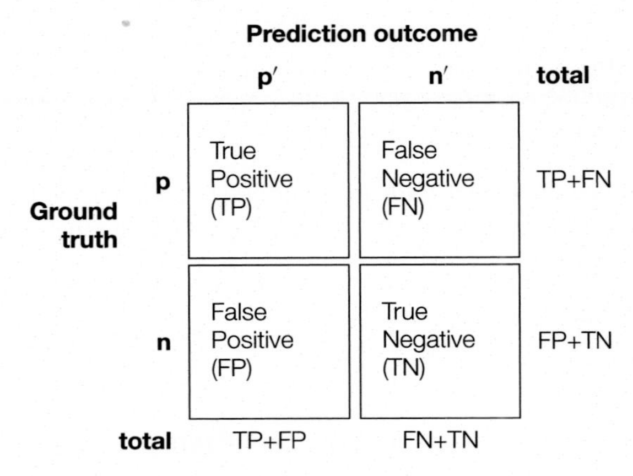

		Prediction outcome		
		p′	n′	total
Ground truth	p	True Positive (TP)	False Negative (FN)	TP+FN
	n	False Positive (FP)	True Negative (TN)	FP+TN
	total	TP+FP	FN+TN	

The F-score is the harmonic mean between precision and recall.

$$F - score = 2 * \frac{Precision * Recall}{Precision + Recall}$$

This means that a high F-score indicates high accuracy and high recall.

An example to show how accuracy alone is in some cases misleading as a measure of performance is given in the case of a dataset in which one class is far more represented than the other. Suppose we have collected data relating to a rare disease and that 98% of the data collected belong to the class of subjects in which the disease is not present and the remaining 2% to the class of subjects with a presence of the disease. A model that classifies all subjects in the dataset as not having the disease will achieve an accuracy of 98%.

4.3.4.2 Evaluation environments

Since the metrics just presented require a comparison between the class predicted by the model to be evaluated and their true class, the ingredients of an evaluation environment are: a set of annotated data and at least one model to be evaluated.

Since the evaluation phase is useful to establish the generalization performances of the model, it is important to keep separate the data used for training from the data used for performance evaluation.

Typically, the initial dataset is partitioned into two parts: One partition will be used as the training dataset and the rest, the test set, for evaluation. This technique, known as split train test, is depicted in Fig. 4.3. The choice of the exact portion to be allocated to one or the other activity depends on the overall size of the dataset.

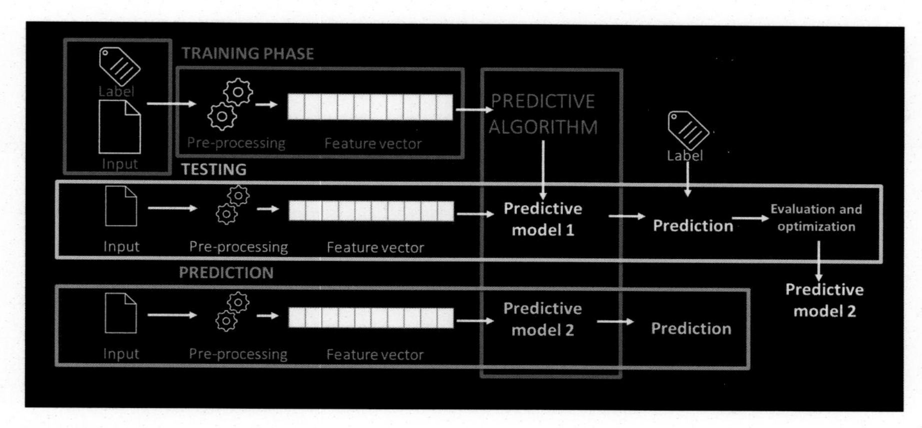

FIGURE 4.3

Example of a train test split evaluation environment.

For not very large datasets, the common choice is to allocate the 70% of the data for training and 30% for evaluation, while, for large datasets, the training set generally encompasses the 90% of the dataset and the remaining 10% is set aside for evaluation.

Since many machine learning models depend on various parameters to be optimized, optimizing the model during the evaluation phase could lead to overfitting on the test set. A good practice is to divide the dataset into three parts: the training set to train the model, the development set for parameters optimization, and the test set for evaluation.

An evaluation technique that allows lowering the dependence of the evaluation of the particular choice of the sets is the k-fold validation: The dataset is divided into k partitions, recursively one of the k partitions is set aside for the evaluation of the model, and the algorithm is trained on the remaining $k-1$ partitions. The performance evaluation will be given by the average of the performances of each model.

4.3.5 Reporting and finalize the model

The last step, i.e., finalizing the model or model deployment, is the planning stage of how to integrate trained, optimized, and evaluated machine learning models into action to support the decision-making system of interest.

However, operationally this step takes on different connotations depending on the specific project, the application domain, the particular decision-making process, and the final user. The deployment step ranges from producing reports and dashboards to developing web applications.

Once the type of model deployment is clear, there are various approaches to finalize it. In particular there exists three different ways to train models in production:

- One-off time—once the model is trained and optimized, it is used as it is and monitored. If the model becomes obsolete, it is updated by training on updated data. Typical formats for the one-off training models are Pick and Joblib;
- Batch training will regenerate the model based on a recurring schedule and make an updated version of the model available based on the latest training.
- Real-time training.

4.4 Languages for data science

The principles, methods and techniques of data science are independent of the type of programming language or the particular platform used for data analysis. In addition, some platforms or libraries that are currently being used to apply data science to specific problems may become obsolete in a few years. The present and the following sections of the chapter therefore do not aim to be an exhaustive and updated review of the most used tools/languages or platforms designed for data science. Rather, the intent is to outline the strengths of leading data science programming languages and to discuss the analysis approaches proposed by specific platforms.

4.4.1 MATLAB®

MATLAB (standing for MATrix LABoratory) is a proprietary programming language developed by MathWorks https://uk.mathworks.com/.

As the name suggests, MATLAB is a high-level language designed to deal with objects represented as a numeric matrix in a fast and efficient way. Besides that, it is possible to integrate MATLAB with various programming languages and systems.

Popular among both engineering and image processing experts, MATLAB offers a range of program extensions and libraries, called toolboxes, integrated within the main MATLAB interface.

The "Statistics and Machine Learning Toolbox" enables statistical functions and makes it easy to build a machine learning application. In particular, the toolbox provides built-in functions for descriptive statistics and for performing exploratory data analysis, different feature transformation, and feature selection algorithms for dimensionality reduction; supervised and unsupervised machine learning algorithms, etc.

Despite the numerous qualities, the use of MATLAB also poses some limitations, the following among them: it has a commercial license; the computational performance is generally lower than with other languages; and the lag in the availability of state-of-the-art algorithms, for example, in the field of deep learning.

4.4.2 Julia

Among the programming languages for scientific computing and data processing, the Julia Programming Language[7] is the most recent, as it has been developed since 2009 by MIT.

Julia is an expressive, high-performance, dynamic and generic programming language. Julia's grammar is as high-level as MATLAB, R, or Python [80]. In addition, Julia combines the high performance of a static programming language with the flexibility of a dynamic programming language, using a just-in-time compiler (JIT) based on low-level virtual machines (LLVM) [81]. This enables reaching a levl of performance comparable to C/C++ by compiling in real time.

Moreover, the built-in primitives for parallel computing better support instruction level parallelism, multithreading, and distributed computing compared to Python. Finally, Julia allows reuse of existing code written in other languages such as R, Python, etc., without the need of a full migration.

Julia's major limitation is that, being a relatively recent language, it has not yet reached the maturity nor the popularity of other languages.

4.4.3 R

R (https://www.r-project.org/) is an open-source language suitable for data analysis and statistical and graphics models, developed in 1995 by Ross Ihaka and Robert Gentleman, [82].

R is widely used for the development of statistical software and data analysis algorithms. Among the reasons for this, it is worth noting that R has explicitly been designed to support statistical analysis, and it offers a very large number of additional packages developed by the community of practitioners, available at the Comprehensive R Archive Network (CRAN) or other popular repositories, such as Bioconductor or GitHub. Moreover, elegant and informative graphics are enhanced by popular visualization packages that enable understand the data in a more efficient and effective way.

However, the computational performance on large dataset may suffer when compared to other languages.

4.4.4 Python

Python (https://www.python.org/) is programming language developed in late 1980s by Guido van Rossum, see [83]. It is an open-source, general purpose, high-level language and, to date, it is certainly the most popular choice among data scientists researchers and practitioners. This is due to its simplicity, the reliability of the written code, the presence of an integrated test framework, and the possibility to integrate

[7] https://julialang.org/.

code inside a notebook, from Jupyter Notebook for sharing code and material. Moreover, as already mentioned, Python is a general-purpose programming language and so allows greater versatility of use.

4.5 Low and no coding tools for data science

In recent years, several companies such as Apple, Microsoft, and Google have made solutions available that enable the development of data science applications with low or no coding effort.

As we have already seen, data science is a field of study in which the data scientist play a central role in creating customized and optimized solutions for specific use cases. Furthermore, as reported by the Venn diagram of data science, nevertheless, low and no-code platforms can help data scientists to automate or speed up repetitive tasks so that their efforts can be focused on other data analysis-related tasks that generally cannot be automated successfully.

4.5.1 WEKA

One of the earliest low coding data science platform is WEKA,[8] standing for Waikato Environment for Knowledge Analysis and developed by the University of Waikato, in New Zealand, see [76], [84].

It is a free software environment written in Java. Weka's popularity lies in the fact that not only does it contain a vast collection of tools for data preprocessing and modeling, along with the most used state-of-the-art data mining algorithms, but it also allows easy access to these functionalities due to a user-friendly graphical interface encouraging use also by non-experts in the Java programming language. On the other hand, for Java developers, it is possible to embed Weka algorithm in a Java application.

4.5.2 KNIME

KNIME,[9] standing for Konstanz Information Miner, see [85], is an open-source platform for data analysis, data reporting and integration, data manipulation, and visualization. KNIME integrates various components of machine learning and data mining through its modular, drag-and-drop data pipeline concept, and allows data to be added in a workflow. The graphical user interface (GUI) facilitates the assembly of nodes for data pre-processing and the ETL process (Extraction, Transformation, Loading), for modeling, data analysis, and visualization.

---

[8] https://www.cs.waikato.ac.nz/ml/weka/downloading.html.

[9] https://www.knime.com/.

4.5.3 PyCaret

PyCaret [86] is a high-level Python library, a wrapper of other popular data science Python libraries such as Scikit-learn, Spacy, etc. The purpose of PyCaret is to implement the training, comparison, evaluation, optimization, interpretation, and distribution of machine learning models with a few lines of code, and consequently to automate and accelerate the cycle of experiments in order to improve productivity.

CHAPTER

5 Deep learning

CONTENTS

Abstract

The set of algorithms designed for modeling a task by using multiple levels of non-linear and nested operation units called artificial neurons falls under the umbrella of deep learning. Deep neural networks have found massive applications in the biomedical as, for instance, biomedical images and gene-expression analysis, genomics, and transcriptomics. This chapter introduces the artificial neuron model and some bases related to feedforward artificial neural networks. We will also discuss one of the central aspects of the popularity and power of deep learning, which is how a new latent representation of the input is learned by the network. Finally, we will provide a taste of the architecture of CNN.

Keywords

Deep learning, Neural networks, Perceptron, Representation, CNN

5.1 Introduction

By the term deep learning we refer to a set of algorithms, also called Deep Neural Networks (DNN), designed for modeling a task by using multiple levels of nonlinear and nested operations [87]. The DNN's capability to model nonlinear tasks has makes them extremely popular in both research and application fields for a wide range of tasks, with particular focus on problems of prediction.

Although deep learning has only recently reached the maturity to enable breakthroughs in numerous application fields, its history is long and many of the fundamental ideas of neural networks date back more than 50 years. In this brief and by no

Artificial Intelligence in Bioinformatics. https://doi.org/10.1016/B978-0-12-822952-1.00014-0

means exhaustive overview of the early days of deep learning, the study by McCulloch and Pitts, who formulated the first neuron model around 1943, must certainly be pointed out [70]. In 1949 Hebb conceptualized the existence of a connection and interaction between neurons as the basis of learning by encouraging the search for computational models inspired by neural activity [88]. Among these the best known is certainly the perceptron, proposed by Rosenblatt in 1958 [89], in which a linear combination of the input is plugged into an activation unit for the calculation of the output.

In the same period, Widrow and Hoff proposed ADALINE [90], an artificial neuron model very similar to Rosenblatt's perceptron, with a linear activation function and a learning rule that seeks to minimize the mean squared error of the training set.

The first embryonic development of the backpropagation model also dates back to around 1960. Another important work of the same years is the one of Hubel and Wiesel [91] on the neurons of the early stages of the visual cortex that at the end of the 70s was taken up by Fukushima et al. [92], who proposed a first example of a multilayer convolutional network, called the neocognitron model, with a learning rule based on the formalization of Hebbs concepts. The neocognitron model and the work of Hubel and Wiesel constitute the foundation of today's CNNs. However, the potential, but above all the limitations, of perceptron models and the skepticism about the possibility of training a "multi-layered" perceptor model appears in the work of Minsky and Papert of 1969 [93]. In 1989 Rumelhart, Hinton, and Williams answered the objections of Minsky and Papert by showing the capabilities of the backpropagation learning algorithm for training deep neural networks [32]. In the same year LeCun presented the architecture of the convolutional neural network (CNN), similar in structure to the neocognitron but trained using the backpropagation algorithm [94]. The backpropagation algorithm is one of the fundamental concepts of deep learning, however it demonstrates the convergence problems for deep networks, due to what is known as the "vanishing gradient" problem [95]. In the late 1990s an improvement in CNN learning was achieved by combining the backpropagation algorithm with the stochastic gradient descent algorithm [96]. If CNNs have been expressly modeled for computer vision tasks, recursive neural networks are an example of neural architectures expressly designed to model sequential data, such as text. The purpose of recurrent neural networks is to model the dependencies between elements of a sequence, which are processed by the network one at a time. For this reason, RNNs do not only provide for a flow of information from input to output (feedforward) as in the case of CNNs, but guarantee the persistence of information by including loops in the network architecture. To overcome the problem of exploding and vanishing gradients which RNNs suffer from, in 1997 the LSTMs were developed [97], a variant of the RNNs that proved capable of modeling long-term dependencies.

From 2011 to today, thanks to the increased computing power due to the implementation of neural networks on graphical processing units and the increasing availability of large datasets on which to train neural network models, deep learning approaches have continued to evolve at a more and more rapid pace.

As a data-intensive domain, deep neural networks have found massive applications in the bioinformatics and biomedical fields. For example, the CNNs represent state-of-the-art approaches for modeling high-dimensional data, as for instance biomedical images and gene-expression analysis. Several applications have been introduced in genomics and transcriptomics.

For further information, readers may refer to the following survey works of deep learning applications in the biomedical field [98–100], while several books can provide interesting details into deep neural networks' architectures and approaches [87,101,102].

The rest of the chapter introduces the fundamental unit of neural architecture, i.e., the artificial neuron model, and some basics related to feedforward artificial shallow neural networks. We will also discuss one of the central aspects of the popularity and power of deep learning, which is which is how a new latent representation of the input is learned by the network. Finally, we provide a taste of the architecture of CNN.

5.2 Introducing basic principles behind deep learning

Before discussing the details of deep learning approaches, in this section we will introduce some basic concepts related to machine learning and data science that have not been explored in previous chapters.

5.2.1 Artificial neuron models

Generally, an artificial neuron is a learning unit inspired by the functioning of the real neurons of the nervous system, and in particular by the first concept of neuron, called the McCulloch–Pitts neuron (MCP). Furthermore, artificial neurons constitute the building block of artificial networks and, consequently, of deep artificial networks. The generic scheme illustrating this learning unit is based on two elements. The first element is a two-phase process described in Fig. 5.1 to determine the map that associates the respective output to an input. In particular, the map is the result of the composition of a linear function, i.e., the weighted sum of the inputs, with an activation function. The figure shows the shapes of the most popular activation functions for these models. The graph of the function in the upper left of the figure shows the Heaviside step function which, together with the linear function, was used in the earliest models of artificial neurons, i.e., the Perceptron, and Adaline. More recently, the logistic or sigmoid function and the tanh have been preferred, while in the last decade it has been shown that the choice of the rectified Linear unit Rectified Linear Unit (RELU) improves the training process of an artificial network. We will see later that, due to its particular properties, the Softmax activation function is very often chosen to generate the output of a deep neural network. The second key element of an artificial neuron is the learning rule which generally aims at optimizing the choice of a vector $\mathbf{w} = [w_0 \dots w_n]$ of weights, also called learning parameters, by optimizing a suitable defined loss function $L(\mathbf{w})$. The loss function provides a measure of how

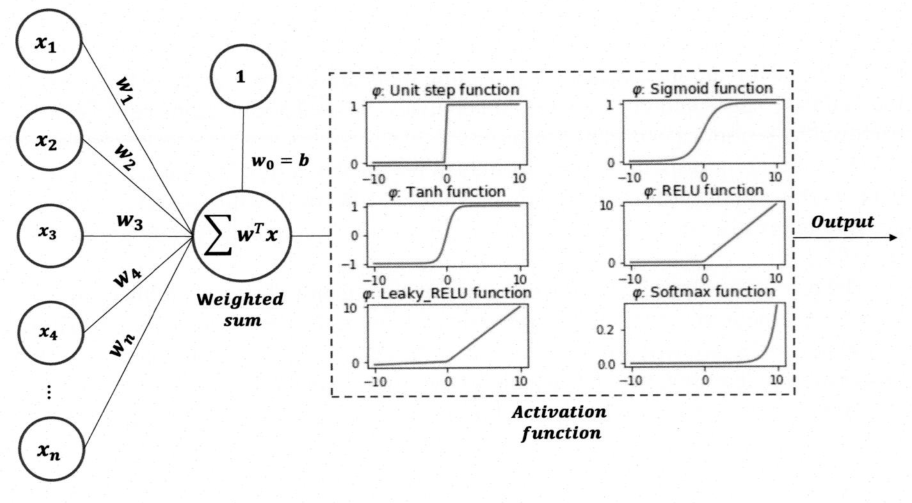

FIGURE 5.1

Artificial neuron and artificial neural networks.

distant is the model output from the true label in the training set, and the optimization is typically performed using the gradient descent algorithm: At each iteration the vector, $\mathbf{w}$ is updated by moving along the opposite direction of the gradient of $\mathbf{w}$ by a learning rate η. Popular alternatives of gradient descent are stochastic and mini-batch gradient descent.

It is worth noting that, except for the identity function, activation functions are generally nonlinear, which is the key point to make a neural network capable of modeling complex nonlinear maps.

An artificial neural network (ANN), or simply neural network (NN), is basically composed by layers of interconnected artificial neurons. The numbers of layers between the input and the output layer are also called hidden layers. In general ANNs with less than one hidden layers have a shallow architecture, while ANNs with more than one hidden layer are termed deep neural networks. The number of neurons composing an hidden layer is known as its width.

In the simplest ANN, the architecture has a feedforward structure, where the connections among successive layers advances from the input to the output, without cycles. Moreover, layers are fully connected, i.e., each neuron of a hidden layer is connected to each neuron of the previous layer.

As in the case of the artificial neuron model, the function that the artificial network model aims to learn depends on the weights of the network connections, i.e., the learning parameters. Training a neural network therefore means determining the values of these parameters that minimize an appropriate loss function. However, in order to train a deep neural network, a learning algorithm such as the stochastic gradient descent is used in combination with another algorithm that provides an efficient method for chain rule-based gradient computation for the derivation of compound functions, i.e., the backpropagation algorithm or simply backpropagation.

5.2.2 Representation learning vs feature engineering

In a simplified way and from a geometric/analytical point of view, the training phase of a machine learning predictive model can be considered as the coding process that, starting from a set of examples and a class of functions, searches for the better approximation of the unknown map that associates the observed outputs to the given inputs.

In the case of predicting continuous values (regression), the sought function approximates the trend of the outputs, given the inputs. In a classification framework, given the input features, the searched map approximates the boundaries of the regions containing the points belonging to the various classes under examination, i.e., the decision boundaries.

Based on the type of the deemed function, machine learning models can be categorized into linear and nonlinear models. If in a classification task the learning set is well-modeled by a linear decision boundary, it is said to be linearly separable. Clearly, most of the real problems exhibit a nonlinear nature. However, it is necessary to observe that the linearity or the nonlinearity of a problem strongly depends on the space in which the problem itself is represented. A famous example is the one reported by Goodfellow et al. [87] and represented in Fig. 5.2. Let suppose we have a two-class problem represented in the Cartesian plane by yellow and blue circles respectively, distributed with respect to each other with concentric circles. According to this representation, a linear classifier would not be able to properly separate the classes. If the same data is represented using polar coordinates, the classes become linearly separable and even training a simple linear algorithm would lead to good results. From this perspective, for example, kernel techniques applied to SVMs or to some techniques for reducing dimensionality, such as PCA kernels and LDA kernels, operate [103].

The previous example shows how choosing the right features space to represent a given problem plays a fundamental role in the construction of a good model. In the classical approaches of data science, the design of the features, i.e., the feature engineering step is based on various experiments that use specific techniques for that precise domain or statistical analysis, etc. Feature engineering approaches produce features that can be interpreted by humans. However, one might wonder whether a human-interpretable representation is best for enabling a machine learning model to learn from data. Furthermore, for some data such as objects recognition from images, it is difficult to know which features should be extracted. Finally, feature engineering is one of the most time- and resource-consuming tasks of the entire data science process.

Unlike classical approaches, deep learning algorithms aim to use machine learning to build an increasingly abstract and nonlinear representation of data when training the predictive model.

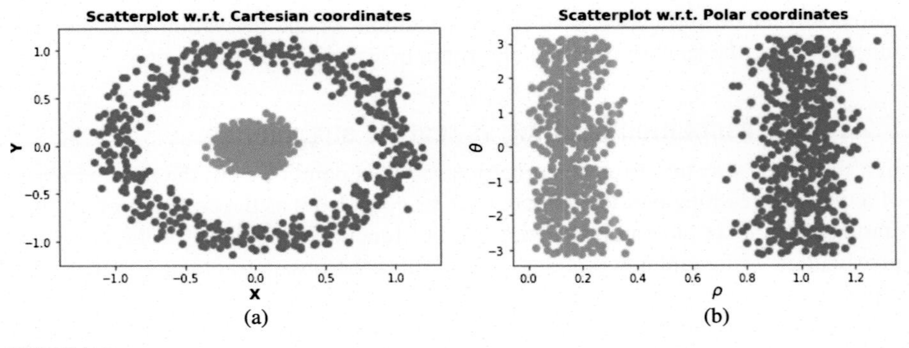

FIGURE 5.2

The importance of representation. In (a) the representation of data using Cartesian coordinates leads to a nonlinear separable problem. In (b) the same data represented using polar coordinates exhibit a linear separable behavior.

To better understand how, let's try to give a trivial example. Suppose we have a dataset containing measurements relating to: familiarity with a given disease (DF), previous pathologies of the patient (PP), results of previous exams (PE), body mass index (BMI), physical activity per day (Act), Smoking (S), Temperature (T), and results of current exams (CE). Now, suppose that, for a suitable choice of the weights w_i, for $i = 1...10$, three new features can be built as linear combinations of the previous ones: Risk Factor (RF), Lifestyle Risk Factor (LRF), and Current Health Condition (CHC). In particular let assume that:

$$RF = w_1 FD + w_2 PP + w_3 PE + w_4 BMI$$

$$LRF = w_5 BMI + w_6 ACT + w_7 S$$

$$CHC = w_8 T + w_9 CE + w_{10} BMI.$$

Then, suppose we were to train an artificial neuron to predict the output class. Fig. 5.3 (a) outlines the described approach with a diagram. The nodes of the leftmost layer represent the initial features, the nodes of the middle layer represent the features we have drawn starting from the initial ones, and the output is calculated from the node of the final layer. Fig. 5.3 (b) represents a feedforward artificial neural network (ANN) with a shallow architecture, while an ANN is usually called a deep neural network if it has more than two hidden layers. It can be noted that the two architectures, in Figs. 5.3 (a) and (b), look very similar.

As shown in the example, the intermediate level of a ANN builds a new representation of the features space starting from the input.

However, compared to the discussed example, a feedforward neural network operates slightly differently. First of all, in a neural network the weights, w_i are not known a priori, but constitute the parameters learned by the model on the basis of

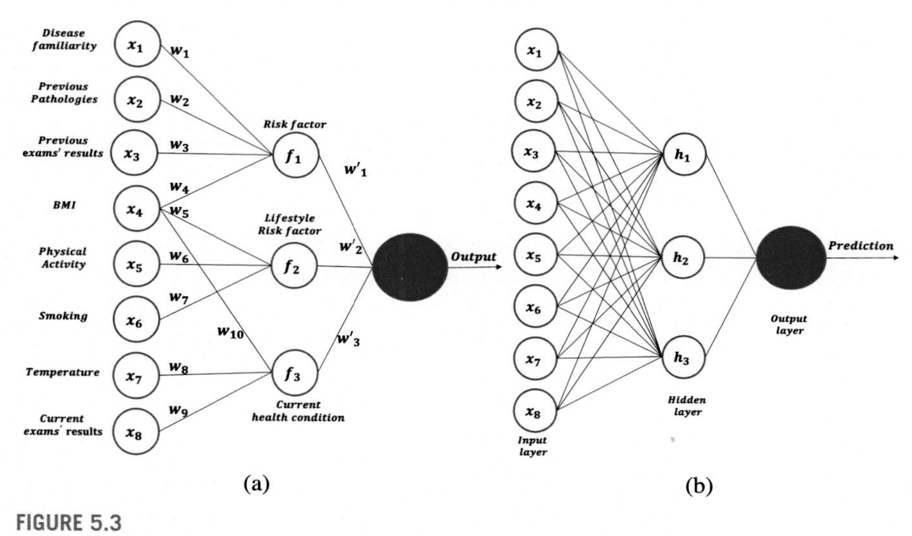

FIGURE 5.3

Artificial neural networks and feature extraction. In (a) the diagram shows the example of hand-designed features extraction. In (b) a shallow ANN is represented.

a learning rule. Consequently, the features extracted in the middle layer of a neural network are not known, nor interpretable, which is why intermediate layers are also called hidden layers.

Furthermore, it is not even possible to know in advance which among the initial nodes will contribute to the construction of the node of the subsequent layer. For this reason each node is fully connected to the previous layer. Finally, in our simple representation, nodes of the intermediate layer are represented by a linear combination of the nodes belonging to the initial layer. In a neural network each node represents an artificial neuron and, as previously discussed, each of the subsequent representations is then built from the previous one through nonlinear transformations.

5.3 Popular deep neural networks architecture

In the previous section we introduced what is meant by feedforward ANNs, and we intuitively discussed how a neural network models nonlinear functions by learning a latent representation in the data being trained. At this point it would make sense to ask what kinds of functions they can model from an artificial feedforward network. The answer to this question is formulated in the universal approximation theorem, which demonstrates that "standard multilayer feedforward networks with as few as a single hidden layer and arbitrary bounded and nonconstant activation function are universal approximators [...] provided only that sufficiently many hidden units are available" [104]. In other words, the theorem establishes that any continuous function on a compact subset of $\mathbb{R}^n$ can be approximated by a shallow feedforward neural network, provided that it has a non-constant activation function and "that sufficiently many hidden units are available".

While this is true for shallow neural networks, there are two reasons for using deep neural networks. The first is that no information is given on how to train this type of network [105]. Additionally, exponential bounded has been provided for the hidden layer's width.

This explosion in terms of the number of nodes of the hidden layer of a shallow neural network can be overcome by increasing the depth of the network. In fact, a universal approximation theorem has recently been proved also for deep feedforwards neural networks with limited width in each layer and having the rectified linear unit as an activation function [106].

5.3.1 Convolutional neural networks

Instead of using fully connected layers, Convolutional Neural Networks (CNNs) are designed on the basis of two fundamental concepts: local connectivity and sharing of parameters. As discussed for the artificial neuron model, CNNs are also inspired by one of the first biological models of the visual cortex, i.e., small regions of neighboring cells are activated in response to specific regions of the visual field, also called local features. To simulate this behavior, a CNN layer exploits three elements: an operation called discrete convolution (although this is different from the standard mathematical convolution operator), a nonlinear activation function, and a pooling layer. In particular, a matrix of weights called the kernel slides across the input. The layer thus obtained is called a feature map because it is composed of neurons which, sharing the same weights, represent a scan of the input in search of neighboring regions in which the same local feature is present. Typically in applications, the discrete convolution operation is followed by a nonlinear activation function like RELU to provide nonlinearity. The pooling layer also acts as a filter shifting across the feature map, but it repeatedly applies the same function which can return the maximum value (max pooling) or the average value of the activity of adjacent neurons.

To extract different feature maps related to different local features, multiple convolutional layers are trained, and the outputs are usually combined through a fully connected layer.

CHAPTER

Explainability of AI methods

6

CONTENTS

Abstract

The popularity of machine learning models and, in particular, deep learning approaches, has made possible the development of applications in a wide range of areas, such as recommendation systems, natural language processing and more, in general, in classification of text, images, speech, etc. However, deep models have the drawback of being highly non-human interpretable. Even though the existing literature does not uniformly agree on a common definition of the term explanation, few practical attempts have been made to build models that justify their actions. This chapter presents an overview of the problem.

Keywords

Explainable models, Artificial intelligence, Machine learning, Deep learning, Bias in AI

6.1 Introduction

It is common knowledge that several among the current state-of-the-art machine learning techniques suffer from three major problems that are particularly limiting in the biomedical domain, namely: generalization problems, coherence issues, and lack of transparency. These three points, are related to a topic that is currently generating particular interest in the scientific community: the interpretability of AI systems.

The nonlinear and nested structure of DNNs that improves accuracy in prediction models does not make it easy understand how, starting from a human-understandable input such as an image or a piece of text, the flow of information reaches a predicted

Artificial Intelligence in Bioinformatics. https://doi.org/10.1016/B978-0-12-822952-1.00015-2

output. In this sense, DNNs are regarded as black boxes. Moreover, since DNNs are generally trained on large amount of data produced as digital traces of real human activities that may contain prejudice or biased, this may lead to predictions that can be in some way discriminative. An example can be found in [107], in which by analyzing 8000 cases using COMPAS, a criminal risk assessment tool, the authors found racial disparities in risk scores, even though data did not include sensitive information such as race.

Another emblematic example concerns the application of machine learning in the health field and, in particular, the prediction of the probability of death in patients with pneumonia to optimize costs by managing patients at low risk as outpatients and admitting high-risk patients [108]. Although the neural network model was shown to be the most accurate, the logistic regression model was instead chosen because it was considered too risky to adopt a model that would not allow medical professionals to understand why and how decisions are made. In fact, a rule-based model extracted the rule that the mortality risk of patients having a history asthma was lower compared to patients having a different medical history. This rule reflected a real pattern, in the sense that asthmatic patients with pneumonia receive more aggressive care compared to other patients, and this common-sense procedure was so effective that it lowered the risk of death in patients with a history of asthma.

Moreover, the need of explanatory systems is required by impending regulations like the General Data Protection Regulation (GDPR), recently adopted by the European Union. The GDPR regulates the collection, storage, and use of personal information, stating that *the data controller shall implement suitable measures to safeguard the data subject's rights and freedoms and legitimate interests, at least the right to obtain human intervention on the part of the controller, to express his or her point of view and to contest the decision*[...]. *Decisions* [. . .] *shall not be based on special categories of personal data unless suitable measures to safeguard the data subject's rights and freedoms and legitimate interests are in place.* So, in order to increase trust in black-box models, the concept of accuracy alone cannot be considered sufficient, and it has to be supplemented with the requirement that the model can be also "explainable".

Currently, there are some proposed explainable models, such as, for example, the Local Interpretable Model-agnostic Explanations (LIME) [109,110], Layer-Wise Relevance Propagation [111], DeepLift [112], and Contextual Explanation Networks (CEN) [113].

6.2 Explainable models in machine learning

Generally speaking, even though the verbs "to interpret" and "to explain" both imply making something clear or understandable, if the former is more related to the idea of an interpreter translating from one language to another one that is more comprehensible to the audience, the latter also implies making something intelligible when is not immediately obvious or entirely known.[1]

In [114], interpretation is defined as *the process of mapping an abstract concept (e.g., a predicted class) into a domain that the human can make sense of*, while an explanation is *the collection of features of the interpretable domain, that have contributed for a given example to produce a decision (e.g., classification or regression).*

In [115], interpretability in machine learning is defined as *the ability to explain or to present in understandable terms to a human*. This definition is also reported in the survey of Guidotti et al. [116], in which an explanation is thought as *an "interface" between humans and a decision maker that is at the same time both an accurate proxy of the decision maker and comprehensible to humans*.

Arguing that existing works in explainable artificial intelligence only take into account what researchers consider should be a good explanation from a computational point of view, the extensive work of Miller [117] gives a definition of explainability by taking into account insights from the social sciences. His major findings are that human explanations tend to be: contrastive, selected among others, that a most-likely explanation does not always imply to be the best possible explanation, and that explanations are social, in the sense that they require a transfer of knowledge. The desirable features of an interpretable model include coherence, generalizability, an improvement in the human-understandability and in the explanatory power of the model [118].

Among state-of-the-art algorithms, there is a small set of recognized interpretable models, such as decision tree or gradient boosted trees and rules-based and linear models. Therefore, one of the simplest way to evaluate interpretability is via a quantifiable proxy [115]: An interpretable algorithm may be seen as an approximation of a model already claimed to be interpretable.

A more general evaluation of the goodness of an interpretable system can depend on the evaluation of the four characteristics previously cited. However, concerning the formalization of suitable measures and methods for evaluating explainable systems, the work in this field is still at an early stage and we believe that further research efforts are needed in this direction.

In general, coherence and fidelity have solely relied on accuracy measures. In particular, when considering the outcome of the original model as ground truth, an external interpreter model is thought to reach high levels of fidelity when it reaches high levels of accuracy.

In systems where the interpretation is provided by an external model with respect to the original model, the interpretability is closely linked to the idea of providing a representation of the original model, especially through the visualization of the most relevant features. In this type of approach, the quality of the produced explanation is deduced through strategies that evaluate the visualization of what the system has learned. An approach of this kind is presented in the work of Samek et al. [119],

[1] https://www.merriam-webster.com/dictionary/interpret.

in which a qualitative measure is proposed to evaluate ordered collections of pixels, such as heatmaps, by region perturbation. This technique has also been used in the context of text classification [120]. For assessing interpretability, to the best of our knowledge, the most frequently used approaches rely on human evaluation and, in particular, on expert evaluation [121–123].

6.2.1 External explainer models

Also, with respect to methods, there is no clear categorization. The taxonomy presented in [118] groups the main approaches into three categories: rule-extraction, attribution, and intrinsic methods.

Rule-extraction methods aim at extracting logical rules, for example, IF–THEN, AND/OR, or M-of-N rules and also decision trees by using the same input on which the DNN is trained and comparing the output of the trained DNN model. In the review of Hailesilassie [124] on Rule Extraction Algorithm for DNNs, three main approaches are highlighted: Decompositional, Pedagogical, and Eclectic. Rule-extraction explaination can be found in the approaches of [125,126].

Attribution or relevance methods are actually the most popular methods and aim to explain the model in terms of how relevant is a feature for the final prediction. Attribution methods are often visualized in terms of heatmaps and performance evaluations usually are related to perturbation, ablation and influence studies. Perturbation strategies are common in so-called Model Agnostic methods, the most popular of which is Local Interpretable Model-agnostic Explanations (LIME) [109,110]. Other popular approaches for explaining deep networks predictions are Layer-Wise Relevance Propagation [111] and DeepLift [112]. There are some conflicting views regarding the lack of sensitivity of local explanation methods [127,128].

Intrinsic methods do not aim to improve the interpretability of a model in a *post-hoc* fashion but by enhancing internal interpretable representations by exploiting methods that are part of the DNN architecture, e.g., by providing an interpretable loss function. With respect to relevance methods, intrinsic methods have the advantage of increasing the fidelity and decreasing the complexity of the resulting explanation.

One of the major issues concerning the current approaches for interpretability is that they address the problem from a technical standpoint, providing explanations that are not meant for the end user [117]. For example, when seeing an image, people recognize a cat, they do not justify the answer by selecting the pixels on which they are based, but perhaps in terms of the ears, nose, tail, etc. Intuitively, this is one reason why Capsule Networks have shown potentially intrinsic explainability properties: They construct a part-whole relationship that can be seen as a relevance path [129]. Since the ultimate goal of an explicable system is to provide explanations to people, it is necessary to take into account also the domain and the levels of knowledge of the people to whom the explanation is to be provided. Interesting approaches in this sense can be found in a few works related to "self-explaining" models.

6.2.2 Self-explaining models

As emphasized in the recent paper of Doran et al. [130], most of the contributes in explainable AI field are focused on enabling explanations of models' decisions, while few attempts are made to build models that "generate" explanations. With the expression "self-explaining", we refer to a model that, in addition to some other task, also generates a line of reasoning in order to explain to the user which decision-making process was followed in terms of input characteristics and by using high-level concepts that are understandable to the user. Since understanding the decision-making process that leads from the input to that specific class should be the final goal of explainable systems, this section is devoted to presenting two examples of approaches to "generating an explanation".

In [121], Hendricks et al. propose a captioning system providing a description that also includes explanations of why the predicted category is the most suitable for that specific input image. Their approach consists of a long-term recurrent convolutional network [131] consisting of a convolutional neural network that extracts highly discriminative image features with the advantage of a compact bilinear classifier and two stacked Long-Short Term Memory (LSTM) networks learning how to generate a description conditioned on the features extracted by the CNN modules. During training, the model receives an instance consisting of an image, a category label, and a ground-truth sentence and is trained to predict each word in a ground-truth sentence, by minimizing a relevance loss function and producing captions describing the input image. In order to provide a sentence generation that is also discriminative on the basis of visual extracted features, another discriminative loss is defined by using a reinforcement learning approach. An aspect that should be highlighted is that, for the evaluation, the authors used both automatic metrics and human evaluations. In particular, automatic metrics were introduced to assess image and class relevance measures for the produced explanation. However, in order to assess if the model provides "explainable justification", a human evaluation was needed.

In [123], a reinforcement learning framework, in which the agent generates explanations for justifying its actions and its strategies, is presented. Reinforcement learning is a technique in which an "agent" learns how to attain a goal in response to the changing state of an "environment" and by improving its actions through experience. In the work of van der Waal et al. [123], the agent answers *contrastive* questions about its actions, i.e., questions related to some contrast case, that can be expressed in the form of "Why X rather than Y?", on the basis of the expected consequences of its policy. In particular, since in contrastive questions facts are posed against counterfactual cases, or foils, in a "why did you choose X instead of Y?" fashion, where X is defined as fact while Y is the foil. Considering an entire learned policy as a fact, a "foil policy" is obtained on the basis of the foil in the user's question.

By learning a state transition model that samples the effect of both the two policies, the expected consequences are modeled as a Markov chain of state visits under both the two policies. The generated explanations, given in terms of state-actions and rewards, are translated into more descriptive state classes and outcomes by following the approaches of [132] and [133]. Also in this paper, in order to assess some model evaluation, human evaluation was considered.

6.3 Application of explainable AI in medicine

In order to develop a Clinical Decision Support Systems (CDSSs) that clinicians can trust, the system needs to ensure that: i) Clinicians understand the predictions (in the sense that predictions have to be consistent with medical knowledge); ii) decision will not negatively affect the patient; iii) the decisions are ethical; iv) the system is optimized on complete objectives; and v) the system is accurate and sensible patient data are protected. Each of the previous requirements is related to one among five scenarios identified in the work of Doshi-Valez and Kim [115], for which it is necessary to request the interpretability of the used model [115,134].

The necessity to build explainable AI systems for the medical domain is well explained in the work of Holzinger et al. [134], in which the authors discussed the insight in developing explainable AI systems for supporting medical decisions. The focus was placed on three data sources: images, omics data, and text. In this work, the authors assessed the need for new strategies to present human-understandable explanations that can also take into account text mining methodologies. Moreover, a possible way to develop explainable AI system for the health-care domain was proposed. The proposal combines: i) A human-in-the-loop machine learning approach, where the model is built continuously and improves over time via feedback and interaction of domain experts; and ii) an interpretable disambiguation system that can provide common sense and human-readable reasons why, in a given context, a given sense was detected [135].

6.3.1 Explainable AI for text mining applications

As stated in the previous section, a basic approach for assessing interpretability is by referring to a quantifiable proxy.

The computational analysis of textual content is also known as text mining. In text mining, learning approaches based on lexicons are considered white-box additive models, and therefore they are trivially interpretable.

The work of Clos et al. [136] proposes a hybrid classification model that performs lexicon-based classification by using a lexicon generated with a learning procedure.

Referring back to the taxonomy presented in the previous section, several attribution approaches have been applied or adapted for explaining text classification models. For example, the LRP model [111] was also applied in order to extract from text which words were most relevant for a CNN's classifier trained on a topic-categorization task [137].

Another attribution approach that was discussed for text classification, is LIME [110]. System evaluation was performed through various human-based assessments. A good critical point was highlighted in the work of Al-Shedivat et al. that compared

LIME with their proposed "self-explaining" approach, i.e., CEN [113], on MNIST and IMDB image datasets. Results showed that when injecting different noise levels in the features, LIME continued to approximate predictions well, while CEN performance was negatively affected from low-quality representations. Results of this evaluation substantially showed that the LIME model can generate misleading explanations. This is another main drawback of "external explanation" models that can encourage further investigation into "self-explaining" models.

To the best of our knowledge, only a few works have considered the production of a textual classifier that can also generate explanations. In [138], the authors proposed an Explicit Factor Model (EFM) based on phrase-level sentiment analysis in order to generate explainable recommendations. As in many previously cited works, also in this case explainability was assessed by showing that the generated recommendations influenced the user's purchasing behavior more.

In [139], the authors propose a modular neural framework to automatically generate interpretable and extractive rationale. An extractive rationale is intended as the concise and sufficient subsets of words extracted from the input text. This subset is required to be interpretable and to represent a suitable approximation of the original output, in the sense that a rationale has to provide nearly the same prediction (target vector) as the original input. A major contribution of this work is that an extractive rationale is considered as a latent variable, generated in an unsupervised fashion, and the generation is incorporated as a part of the overall learning process leading to a justified neural network prediction. A minor point is that an interpretable extractive rationale is defined as a coherent and concise subset of text, but interpretability is solely assessed as a consequence of the extraction process.

Consistency and interpretability are not quantitatively evaluated and, under the assumption that better rationales achieve higher performance, evaluations of the proposed framework are based on accuracy measures such as Mean Average Precision (MAP) for a similar text retrieval task and Mean Squared Error (MSE) on the test set for the multi-aspect sentiment-prediction task.

CHAPTER

Intelligent agents

7

CONTENTS

Abstract

Continuous advances in biological sequences, transcriptional and structural data, interactions, and genetics have led to the production of massive amounts of data that require the adoption of applications and integrated tools for unsupervised automated analysis of this continuous growing flow of biological data. Intelligent agents can support bioinformaticians in simulating and modeling biological systems, as well as supporting the automation of information gathering and information analysis processes. In this sense, an agent is an intelligent, flexible, and autonomous software tool with problem-solving skills. Agents are high-level software abstractions providing a practical and powerful way to describe a complex software entity regarding its behavior within a contextual computational environment. Thus, agents can perform actions without user mediation by maintaining a description of their processing state and the outer environment in which they are situated.

This chapter introduces how to define an intelligent agent according to the FIPA (Percepts, Actions, Goals, and Environment) specifications and how to provide it with the necessary independence to deal with various problems through perception and action.

Keywords

Agents, Multi-agent systems, Biological systems simulation, JADE, JADEX, MESA

7.1 Introduction

The term Intelligent Agent (IA) defines a software entity contained within an environment and endowed with the necessary autonomy to determine, through perception and reasoning, without supervision and in an autonomous way, the consequent actions to be taken [140]. Fig. 7.1 is a schematic rendition representing a generic IA.

Artificial Intelligence in Bioinformatics. https://doi.org/10.1016/B978-0-12-822952-1.00016-4

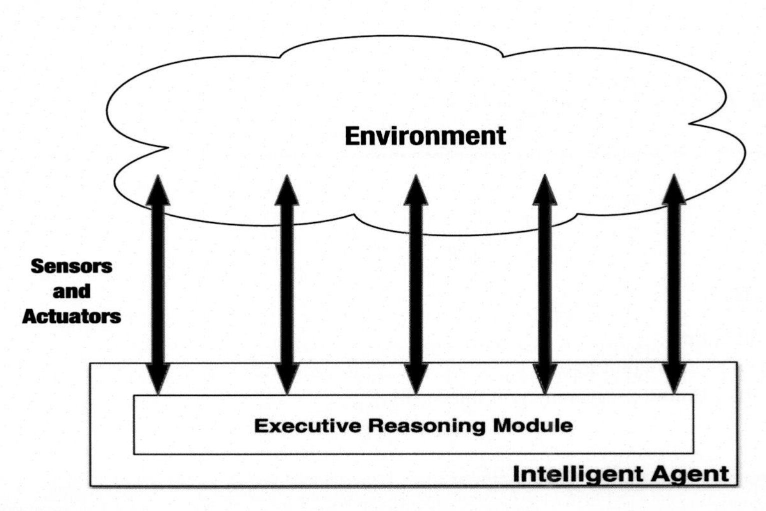

FIGURE 7.1

Representation of the IA's interaction with the external environment and consequent reasoning.

An IA is defined as *rational* if, and only if, it is capable of pursuing specific objectives based on the available information in order to carry out complex activities in the environment surrounding it. A *rational agent* can act in the optimal way to accomplish the assigned task. It must maximize the expected values of the performance measures, given the existing collected data sequence and the acquired knowledge. The intelligence possessed by a rational agent must be action-oriented, also known as practical intelligence. An agent's rationality depends on the current computational resources, whose limits (time, costs, memory, etc.) do enable attainment of perfect rationality. The agent's rationality must allow it to know the surrounding environment by employing special sensors [141], enabling it to update its knowledge based on the newly collected information.

A rational IA is a computational entity, i.e., a software program or a robot equipped with an algorithm for processing input data in order to transform signals into decisions or actions, for example, detecting anomalies in a heartbeat. In general terms, the IA's action choice at any moment is due to a series of events detected up to that moment. For instance, let's consider swarms' behavior. As stated in [142], swarm behavior is the collective motion of a large number of self-propelled entities. Nowadays, the term swarm is applied to inanimate entities if they exhibit similar behaviors, like robots swarm or stars swarm. The most significant advantage of swarm-based solutions consists of the simplicity of each entity's rules, ensuring a resulting global, complex, and intelligent behavior not involving any central coordination. Swarm behavior was first realized on a computer in 1986 with a simple program simulating agents that are allowed to move according to a set of basic rules designed to imitate birds' flocking behavior [143]. This result can be reached using a suitable mathematical model known as agent's behavior mapping functions to produce correspondence between any percept sequences and one or more specific action(s) to be completed.

$$\beta : P_{\mathbb{H}} \rightarrow A_{\mathbb{S}} \tag{7.1}$$

In Eq. (7.1), β is the behavior's mapping function, $P_{\mathbb{H}}$ represents the precept sequence history, and $A_{\mathbb{S}}$ is the action or the actions' sequence.

The mapping function is implemented like a table, made up of columns related to input and output, while the rows represent the various mapping modes among the collected data and the consequent actions. The IA is the software program that has to implement the function $\beta()$ and the mapping table. In this manner, it is possible to create a software agent, e.g., an autonomous program that can relate perceptions and actions. The appropriate autonomous program that implements the mapping function concretely converts the input into the output data.

Thus, a software agent can be defined as the union between the sensors and actuators, and this is usually represented with the following assertion:

$$\mathbb{A}gent = (\mathbb{S}ensors \;\&\; \mathbb{A}ctuators) \cup \mathbb{S}oftware,$$

i.e., the hardware and the software that drives the hardware to accomplish the goals.

The intelligent agent description takes place through the environment of use, perceptions used, objectives pursued, and actions performed, as indicated by the PAGE [144] specifications (Percepts, Actions, Goals, and Environment) used to define agents and indicate their characteristics.

To this end, various different types of agents have been defined with various characteristics and increasing complexity: *Simple Reflex Agents*, *Model-Based Reflex Agents*, *Goal-Driven Agents*, *Utility Based Agents*, and *Learning Agents* that will be discussed in more detail in the next chapter.

Each model can extend the previous models, including their characteristics, to create behaviors that can achieve increasingly complex forms of intelligence.

The notion of intelligent agents in bioinformatics suggests designing domain-aware IAs able to deal with biological information. Thus agents in bioinformatics lead to the creation of customized agent-based systems, tools, and languages for modeling the biological processes themselves. An example to illustrate the IA's role in bioinformatics [145,146] could concern a community of agents that interact with each other to automate omics data annotation and enrichment and prioritization processes. Generally, a single program cannot perform a complete analysis, starting with collecting omics data from multiple remote databases and available in different formats. That requires including further handling steps such as harmonization, preprocessing, annotation, enrichment, and, finally, knowledge extraction. Thus omics data analysis must be done by integrating several distinct computational frameworks, creating a software pipeline in a sophisticated manner, as shown in Fig. 7.2(a). The multi-agents approach represents a good solution to automatize the single steps of the omics data analysis pipeline, as shown in Fig. 7.2(b).

In a multi-agent environment, several types of agents can independently and without external supervision collect data from the central databases and perform data harmonization, preprocessing, and knowledge extraction, in a more effective and flexible manner through the existing tools, without the necessity of offering new methods to perform these tasks (Fig. 7.2(b)). In recent years, several initiatives involving the use of IA in bioinformatics have been emerging. In particulars, IAs to analyze gene expression and Single Nucleotide Polymorphisms (SNPs) data sets. IA can be useful to understand how SNPs' presence/absence or up-regulated genes may affect the development or progression of a disease like cancer, diabetes, and Alzheimer's. The use of agents to analyze gene expression or SNP data sets could help understand the links between genes and their association with diseases by employing pathway enrichment to explain the biological function affected by the over-expressed genes [147].

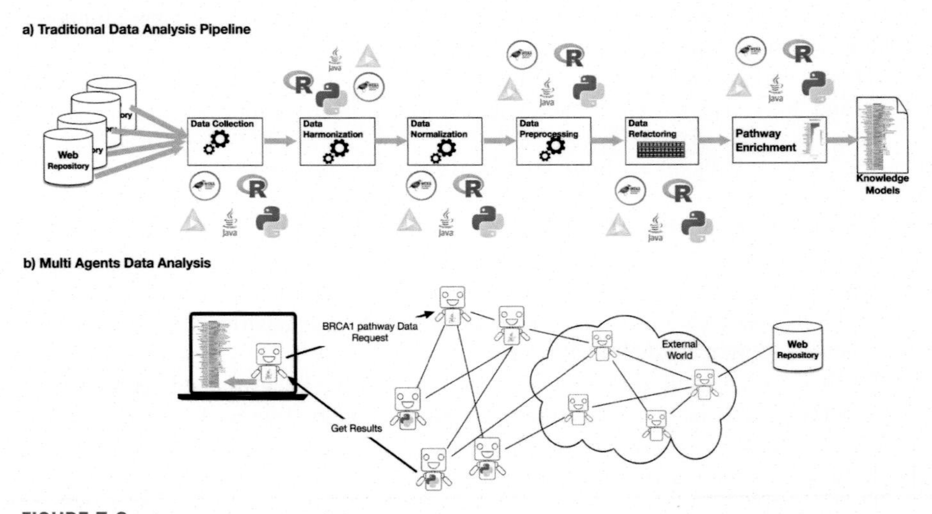

FIGURE 7.2

Sub-figure (a) represents a common analysis pipeline used in bioinformatics. It is worthy to note that each analysis step has to be explicitly executed by the users each time that it is required, using the most suitable programming or script language. Conversely, in sub-figure (b), after the creation and execution by one or more IAs, they can independently accomplish the goals for which they were developed.

Predicting protein 3D structure is a complex task because the underlying process involves biological, chemical, and physical interactions. A simplified task is to predict the secondary structure, i.e., the peptide chain's local conformation projected into a one-dimensional sequence. Despite this simplification, in order to predict the secondary structure it is necessary to use Neural Networks (NNs). NNs have been widely applied to this task due to their many successful secondary structure prediction thanks to their ability to find patterns without the need for predetermined models or known mechanisms. IA can also be used to predict the secondary structure of a protein. In [148], the authors present an agent-based framework for ab-initio simulations, composed of various levels of agents, stratifying the agents in three layers according to their knowledge and their power. The first layer contains the agents designed to explore the states' space, the second layer deals with agents implementing the global strategies, the last one holds the cooperation agents. The amino acids in the protein are modeled as an independent agent that can explore the configuration space. This is accomplished mainly by letting these agents interact and exchange information with their spatial neighbors, as depicted in Fig. 7.3.

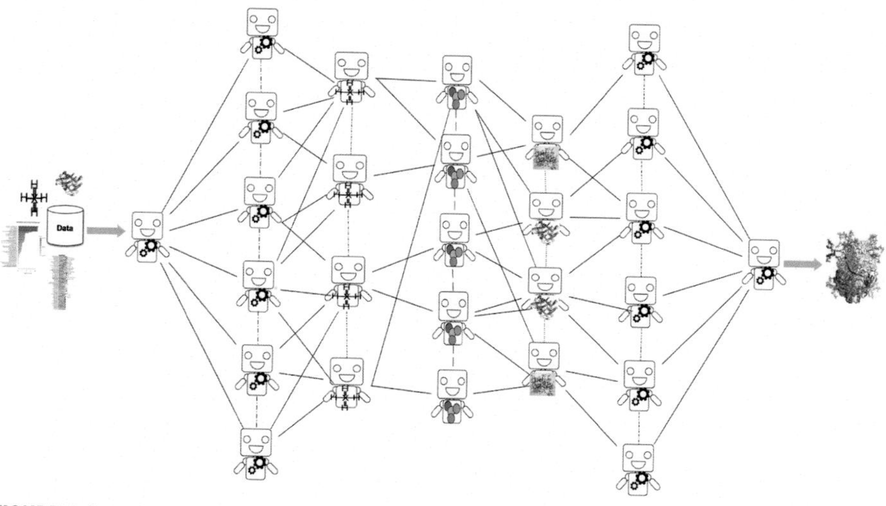

FIGURE 7.3

A multi-intelligent agent collaboration system. Various agents interact and exchange information with their neighbors so that they can effectively and without user supervision figure out the secondary structure of an unknown protein.

IAs have also been applied in health-care services to improve medical diagnosis. In medical diagnosis, the intelligent agents use the patient's information to decide the best sequence actions. In this scenario, the patient is considered the environment, whereas the network card is the input sensor that receives data on the patient's symptoms from the internet connection. Finally, medical attention is given through actuators such as treatment updates.

7.2 Types of intelligent agents

Intelligent agents can be of multiple types and are classified into agents and subagents according to their peculiarities, always keeping in mind their PAGE composition. For example, physical agents can be made with sensors and actuators or temporal agents that use time-based information to provide instructions and collect inputs to change their behavior based on a perceived stimulus at any moment.

We can classify agents into five types:

- **Simple Reflex Agents.** This represents the most straightforward scheme because it allows agents to react only to external stimuli received through the sensors. Perception is the input sent to the agent at a given moment, so there is no internal memory of what is happening outside the environment. Actions are determined based on the current perception through a set of rules, e.g., *condition–action* type, neglecting the past sequence of perceptions. Fig. 7.4 presents the components and the flow of information of this type of agent. Although such an agent kind can be implemented quickly and efficiently, its range of applicability is narrow.

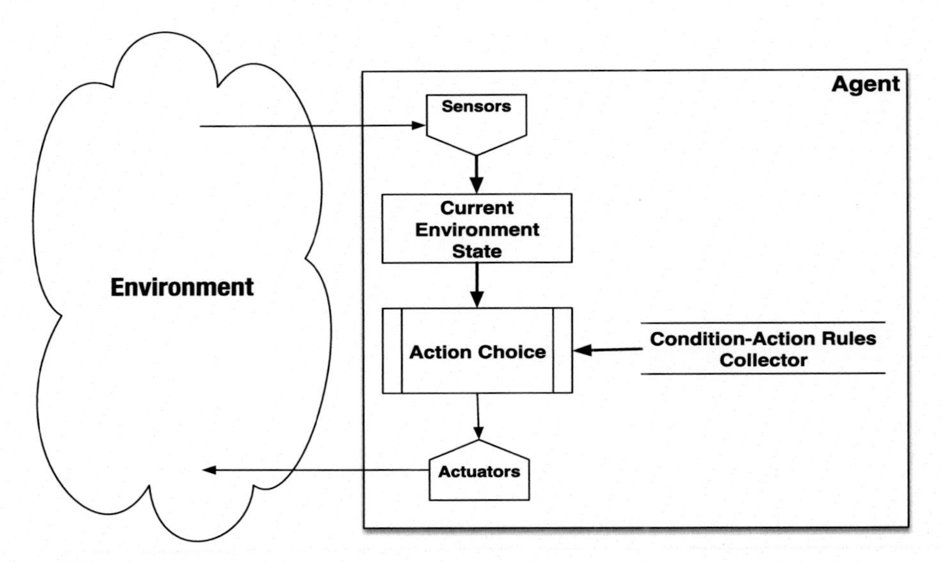

FIGURE 7.4

Representation of a simple reflex IA. This is the simplest schema because the agent implementing such a model can react only to the received external stimuli.

- **Model-Based Reflex Agents.** This type of agent introduces the concept of *internal state* in order to select the action to perform. This extension is necessary when the correct action to perform cannot be decided based only on the current perception, as happens in the simple reflex agent. The internal state (e.g., how the agent's actions affect the world) is constructed from the perceived sequence, along with the current perception (e.g., how the external environment evolves independently of the agent), making possible to determine the agent's condition. Both establish a representation of the outside world. The internal state is updated with each perception. Briefly, the agent creates and modifies its model, an expression of the external environment. Since the agent knows the current condition of the external environment, the choice of an action occurs according to the condition–action

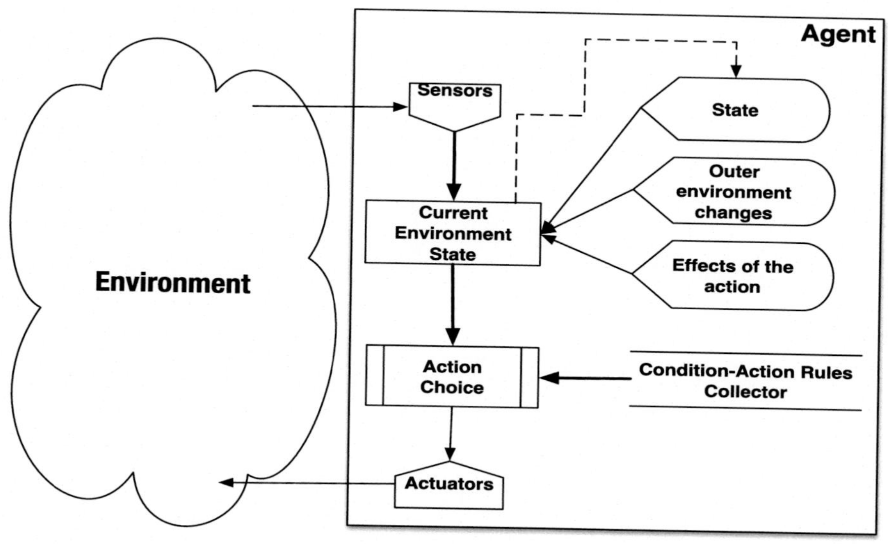

FIGURE 7.5

Representation of a model-based reflex IA. This schema introduces the possibility for the agents that implement it to have the ability to build a model of the external environment, handling the sequences of perceived external stimuli.

rules. Fig. 7.5 shows the components and the flow of information in this type of agent.

- **Goal-Driven Agents.** This type of agent can manage explicit information regarding the goal to be achieved through action planning. Knowing in some cases the current state of the outer environment, it is not enough to conclude what to do. Thus, the right decision depends on the goal pursued by the agent. In this case the goal represents a collection of information describing eligible situations. The choice of actions takes place by combining the internal state, the aim to be pursued, and the evaluation of the effects of the possible actions that can be taken. In this way, to create flexibility for changes, how the environment will change by acting on its model is estimated. Fig. 7.6 shows the components and the flow of information in this type of agent.
- **Utility-based agents.** Utility expresses the degree of reward for an agent when a specific state in the external environment has been reached. The utility function assigns to each state a real value that quantifies the degree of fulfillment reached by the agent. This type of agent is advantageous when multiple conflicting goals cannot be achieved together (for example, privacy and data access) because it makes possible selecting which one to privilege. It is also useful when several goals are achievable but none are feasible with certainty because it enables comparing the

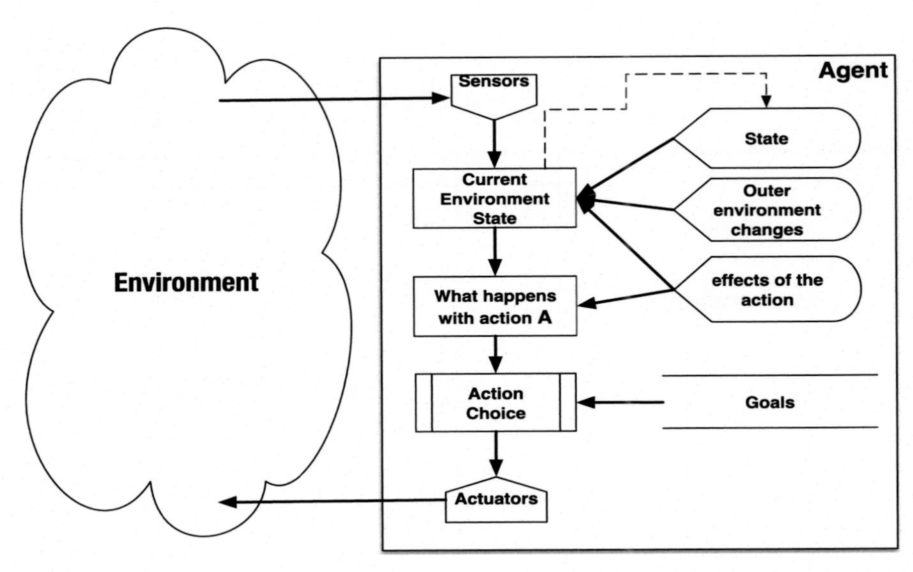

FIGURE 7.6

Representation of a goal-based IA. This schema allows agents to explicitly schedule the actions to reach the planned goal.

probabilities of success and the importance of the goals. Fig. 7.7 presents the components and the flow of information in this type of agent.

- **Learning agents.** Programming of agents is a computationally difficult operation. Artificial intelligence is used to ensure that the system selects different elements to implement the corresponding action. This type of agent uses an artificial intelligence module consisting of the following sub-modules: learning module, execution module, critical analysis module, and problem generator module. In this configuration, the internal learning module improves the model from time to time through learning. Based on the information received, it decides whether and how to modify the execution model's knowledge to achieve better results in the future, while determining where it can be extended or improved. The execution module deals with the choice of actions based on initial knowledge and starts to operate on the external environment by choosing some actions. The performance analysis module communicates to the learning module how the agent behaves concerning a predetermined performance threshold and classifies the incoming perceptions as rewards or penalties achieved by the agent. The problem generator suggests new actions that allow the executive module to perform better than using the previously available learned actions. Perceptions do not indicate the agent's success. In this way, the actions are performed independently, and the agent can adapt to the most diverse evolving conditions. Fig. 7.8 shows the components and the flow of information in this type of agent.

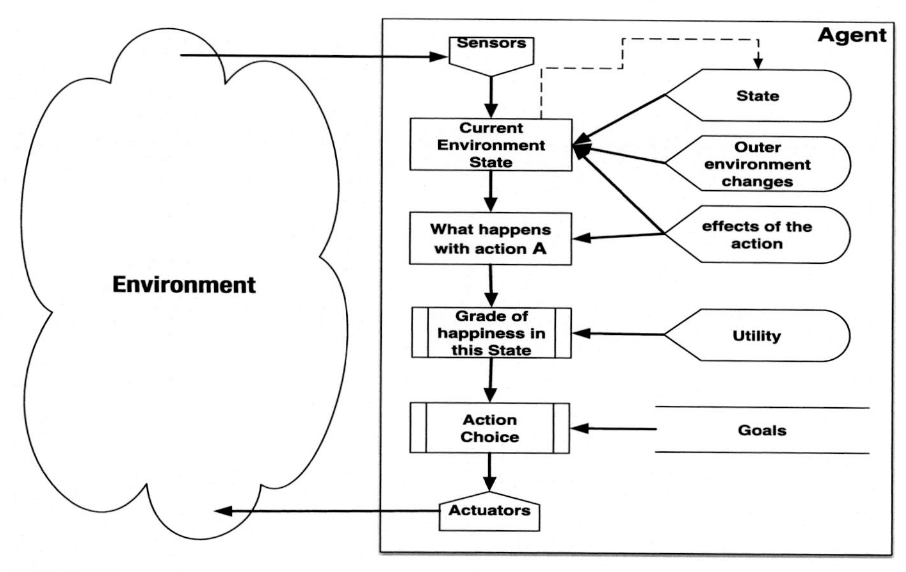

FIGURE 7.7

Representation of an utility-based IA. This schema introduces the concept of rewarding for an agent based on the achievement of a possible goal.

In addition, because each agent operates in an environment, the analysis of the environment in which to place the agent is fundamental so that we can classify the environments and their characteristics as:

- Accessible or inaccessible whether the sensors give the agent access to the complete state of the environment or not. We can say that the environment is accessible to the agent if the sensors can percept relevant stimuli and react accordingly. For agents acting in an accessible environment, it is unnecessary to keep track of the external stimuli.
- Deterministic or nondeterministic, whether the next agent's state is uniquely determined by the current one and by the agent's selected actions. Sometimes, it is better to classify an environment as deterministic or nondeterministic based on the agent's point of view. Inaccessible environments may seem nondeterministic to the agent since in complex environments it is challenging to be aware of all the unavailable aspects.
- Episodic or nonepisodic. Any episode is given by the agent's perceptions and consequent actions. In a single case, subsequent episodes do not depend on what actions happened in the previous episodes. Episodic environments are much simpler because the agent does not need to think about the future.

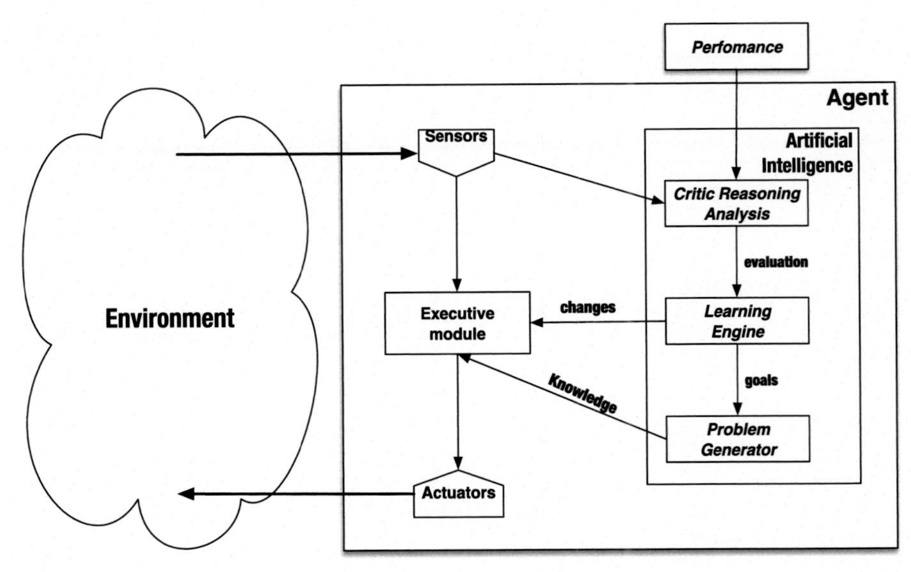

FIGURE 7.8

Representation of a learning IA. This module integrates the artificial intelligence module, allowing each agent implementing it to improve its actions by exploiting new and previously accumulated knowledge.

- Static or dynamic. An environment is dynamic if it can change while the agent is deliberating; otherwise, it is static. Static environments are simpler because agents do not need to monitor the external environment while deciding an action, nor should they worry about the passing of time.
- Discrete or continuous. An environment is discrete if it is described by a limited number of perceptions and consequent actions. For example, a discrete environment encloses the gene name translation within the protein identifier. In contrast, an environment is continuous based on whether its properties vary continuously, for example, the acquisition of a microarray of data images.

An example of a bioinformatic environment would be medical diagnosis systems. The agents operating in this environment are called medicine diagnostic agents. The diagnostic agents interact with the external environment, including patients, hospital facilities, staff through sensors, e.g., a keyboard to enter the symptoms, findings, patient's answers, or a body sensor peacemaker to monitor the patients' heart rate after heart surgery. The actions taken are meant to maximize performance; possible performance measures comprise costs, lawsuits, and health. The actions are accomplished through actuators or a screen display that shows the actions' results, e.g., diagnoses, treatments, and referrals. Thus, different environments require very different agents whose relationship defines the type of agent to use.

7.3 Agent-oriented programming frameworks

Albeit an intelligent agent can be developed using any programming language, usually object-oriented (OOP) languages, such as Java, C++, and Python, are not recommended. The use of general-purpose programming languages to develop agents requires additional efforts to manually handle particular aspects of the agent specification, such as explicit control of the network and communication layers in order to allow agents to communicate to accomplish their tasks. Consequently, the concept of agent-oriented programming (AOP) was introduced in [149], which can be considered a specialization of object-oriented programming. The AOP model provides the necessary structures and paradigms to design and implement agents using a set of high-level programming constructs. These programming constructs simplify the implementation of the agents' essential functions such as cooperation, communication, and synchronization. Thus, AOP leaves the programmer free to focus only on the definition of the agent's intelligence in order to address the problem to solve. There are many AOP frameworks, also named agent platforms, among which are Java Agent DEvelopment Framework (JADE) [150], JADEX [151] (JADE extension), and MESA [152]. Appendices A and B present some examples of Python and Java codes related to the three frameworks.

PART

2

Artificial intelligence: bioinformatics

Part 2 outline

Bioinformatics involves biology, computer science, mathematics, statistics, and other disciplines, with the aim to represent, analyze, and interpret biological data and processes. It uses several computer science data abstractions for modeling biological data and applies several algorithms for extracting knowledge from such biological data. The second part of the book deals with basic concepts and algorithms of bioinformatics, including sequence and structure analysis, main omics sciences, biological network analysis, and pathway analysis. Emergent disciplines that connect bioinformatics with artificial intelligence, such as ontologies, integrative bioinformatics, and knowledge extraction from texts, such as biomedical literature or medical reports, are also described.

Chapter 8 introduces sequence analysis, a keystone of traditional bioinformatics (sequence alignment algorithms are central to compare DNA or amino acids sequences), which is attracting more interest due to the development on next-generation sequencing which generates huge volumes of sequence data.

Chapter 9 introduces structure analysis, another keystone of traditional bioinformatics. This chapter focuses on algorithms and methods for the prediction of secondary and tertiary structures of proteins, which have been improved in recent years due to the application of artificial intelligence.

Chapter 10 introduces omics sciences, which perform an important role in revolutionizing the bioinformatics world. The chapter describes the main omics sciences, such as genomics, trascriptomics, epigenomics, proteomics, and metabolomics and relevant data analysis methods.

Chapter 11 introduces ontologies, a computer science approach to building knowledge bases for bioinformatics. The chapter introduces the primary biomedical ontologies, including gene ontology, and reports on the primary data analysis methods realizable through ontologies, such as molecular data annotation, semantic similarity measures, and functional enrichment analysis.

Chapter 12 introduces integrative bioinformatics, a novel discipline that aims to improve the analysis of biological data through a systematic integration of several data sources. The chapter discusses the integration and analysis of heterogeneous omics data, and describes the main publicly available data sources that enable researchers to study several diseases by leveraging big data sets of genomic and proteomic data.

Chapter 13 introduces biological networks, such as protein interaction networks and the key methods developed for motif discovery and network comparison.

Chapter 14 introduces biological pathways, a special case of biological networks, that are central to capturing the dynamics of biological networks.

Chapter 15 introduces an emerging approach to study biological and clinical data based on text mining, i.e., knowledge extraction from written texts.

Chapter 16 recounts the evolution of bioinformatics and summarizes the issues and challenges of using artificial intelligence in bioinformatics that may prevent its use in real applications.

Finally, Appendix A and Appendix B report some source codes, respectively, in Python and Java programming languages, which are employed throughout the book, to explain the most important bioinformatics and artificial intelligence algorithms.

CHAPTER

Sequence analysis

8

CONTENTS

Abstract

DNA and RNA molecules are represented as long sequences of symbols, letters, and dashes called strings. Essentially, DNA and RNA sequences are like strings meaning ordered sequences of symbols. In fact, the basic representation of nucleic acids is simply a letter in the alphabets $\{\mathbb{A}, \mathbb{C}, \mathbb{G}, \mathbb{T}, \mathbb{U}\}$. Also, the term sequence is defined as a synonym for string and is used interchangeably in various contexts. A string is an ordered sequence of symbols: The interpretation of a string conveys its meaning. More formally, the elements of a string are drawn from an alphabet. An alphabet Σ is a nonempty set of elements called symbols or letters. A string s is a finite ordered sequence of symbols defined over Σ. Studying and analyzing strings or sequences constitutes a crucial task in various contexts, ranging from computer science fields such as data mining to biological contexts such as proteins or DNA alignments. This chapter presents and examines some well-known string similarity methods available in the literature.

Keywords

Sequences, Sequences alignment, DNA and RNA sequences, Edit distance, Dynamic programming

Artificial Intelligence in Bioinformatics. https://doi.org/10.1016/B978-0-12-822952-1.00018-8

8.1 Introduction

Living organisms contain the essential instructions for life within their genetic material or **genome**. The genomes of cells comprise the deoxyribose nucleic acid (DNA), while the viral genome is composed of ribonucleic acid (RNA). The building blocks of DNA and RNA sequence are the purines and pyrimidine bases. Purine bases include the nucleotides Adenine (A) and Guanine (G), occurring both in DNA and RNA. Pyrimidine bases include Cytosine (C), Thymine (T), and Uracil (U). Cytosine and thymine occur in DNA, whereas uracil replaces thymine in RNA. Nucleotide bases are also known as the genetic alphabets $\aleph = \{A, C, G, T, U\}$. The base pairing between purine and pyrimidine is complementary, which means that: In DNA, A pairs with T, and G pairs with C. In RNA, A pairs with U, and G pairs with C. The order of these nucleotide bases determines the meaning of the information encoded into the DNA and RNA molecules. Thus, DNA and RNA molecules are represented as long sequences of symbols, letters, and dashes called strings.

Essentially, DNA and RNA sequences are like strings, e.g., sequences of ordered symbols. Before giving a precise and formal definition of a string, let us first concentrate on some examples. Throughout this entire chapter, we distinguish strings by using a monospaced font `such as this one`. Also, the term sequence is defined as a synonym for string and is used interchangeably in various contexts. A string is an ordered sequence of symbols: The interpretation of a string conveys its meaning.

An alphabet Σ is a nonempty set of elements called symbols or letters. A string s is an (finite) ordered sequence of symbols defined over Σ, i.e., s consists of symbols from Σ. A string is represented by the juxtaposition of the symbols it is composed of. As an example, suppose $\Sigma = \{\texttt{A}, \texttt{C}, \texttt{G}, \texttt{T}\}$ is the DNA nucleotide alphabet, then $s = \texttt{AGGTCA}$ is a string defined over Σ. In some cases, it is useful to refer to the set of symbols occurring in a string s, denoted by $\text{alph}(s)$. The number of symbols in a string s indicates its length, and we denote it as $|s|$. If $|s| = 0$, then s is an empty string that we denote as $s = \epsilon$. Usually, a string allows its symbols to be referred to by an index. In particular, we denote as $s[i]$, for $i = 1, 2 \ldots, |s|$, the symbol at index i of s. If $s \neq \epsilon$, we write s as the juxtaposition of its symbols $s = s[1]s[2] \ldots s[|s|]$. It is important to point out the elementary definition of identity between two strings s and t, and formally $s = t$ if, and only if, $|s| = |t|$ and $s[i] = t[i]$, for $i = 1, 2, \ldots, |s|$.

In particular, a substring of s is a string that can be derived from s by deleting some symbols at the beginning and/or at the end of s. Formally, a string p is a substring of s if $s = upw$, where u and w are two strings. If $u = \epsilon$, we call p a prefix of s, whereas if $w = \epsilon$, we call p a suffix of s. We can also indicate a substring of s as a sequence of contiguous indexes in s, and we denote it by $s[i..j] = s[i] \ldots s[j]$, with $i \leq j$, indicating the substring of s starting at index i and ending at index j. A substring p of s naturally occurs in the latter. Formally, let p be a string $(p \neq \epsilon)$ and let s be a string: we say that p occurs in s if p is a substring of s.

A subsequence of s is a string that can be derived from s by deleting some symbols, without changing the order of the other symbols. In particular, a subsequence of s is a sequence of (possibly) not contiguous indexes in s. Formally, let s be a string, then p is a subsequence of s if exists a sequence of indexes $i_1, i_2, \ldots, i_k$, with $i_1 < i_2 < \cdots < i_k$ and $k = |p|$, such that $p[j] = s[i_j]$ for $j = 1, 2, \ldots, k$. Also in this case, having two strings p and s, we say that p occurs in s if p is a subsequence of s.

As an example, consider the string $s =$ `abbdecbba`. A substring of s is $p =$ `bbd`, having a single occurrence in s starting at index 2, while $q =$ `adb` is a subsequence of s, having two distinct occurrences in s, consisting of the indexes $1, 4, 7$ and $1, 4, 8$ respectively.

8.2 String similarity methods

Strings play a fundamental role in computer science since data is codified into strings; various examples are time series, access logs, astronomical data, financial data and so on. The academic research literature has treated strings for many years, formulating a fundamental theory that plays a crucial role in almost every, but not limited to, field of computer science.

Studying and analyzing strings constitute crucial tasks in various contexts, ranging from computer science fields such as data mining to biological contexts such as protein alignments. As a matter of fact, strings present a neutral representation for several kinds of data. In Fig. 8.1 we show two examples of data representation by strings. In particular, Fig. 8.1 (left) shows a symbolic representation of a time series. This kind of representation can be obtained by a discretization algorithm, such as SAX [153], which takes in the data points as input and discretizes them into a sequence of symbols. In contrast, Fig. 8.1 (right) shows a short part of a DNA/gene represented as sequences of nucleotides.

A central task of many studies and applications is that of computing the similarity between two strings. Given two or more strings, a few interesting questions arise, such as: "are these strings related?", and "if they are related, can we measure this relatedness?". Measuring the similarity between time series can be useful for identifying temporal shifts or data spikes that might indicate some unexpected behavior [154]. Instead, measuring the similarity between protein sequences and, more in general, the problem of protein alignment is at the basis of many bioinformatics applications, thanks to its key role in evolutionary, molecular and development biology [155]. In general, defining some sort of measure of similarity between two entities is the foundation on which to build several tasks. A classical example of similarity among objects is clustering. Clustering is the process of finding groups of similar points in a data set [156]. These groups are called clusters, and each cluster represents a set containing similar data points w.r.t. a certain definition of similarity.

In the case of strings, several definitions of similarity between them have been proposed in the literature, ranging from classical to more elaborated ones. Strings introduced the notion of distance, i.e., a way to measure how different two strings are. Usually, two strings are similar if their distance is low, whereas they are dissimilar if their distance is high. Various methods, such as the *edit distance* or the *longest common subsequence*, are based on such a notion and all define some sort of dis-

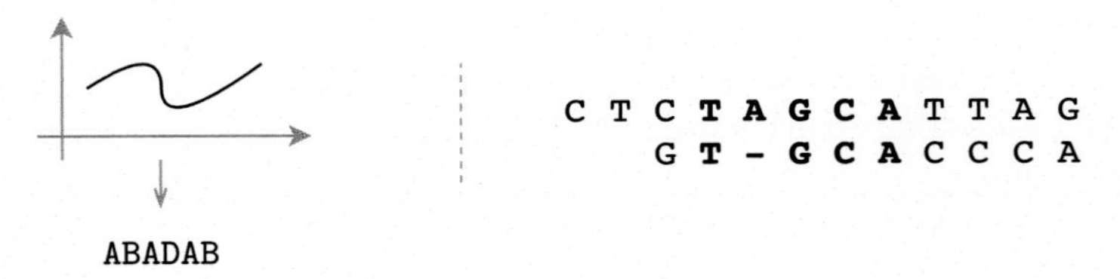

FIGURE 8.1

Example of string representation of different entities, such as time series (left) and gene entities for alignment (right).

tance. Also, they are all based on a particular natural assumption: Identical symbols among strings represent identical information, whereas different symbols introduce, in a way or another, some form of differentiation. This assumption generally holds, although it appears to be extremely reductive in different scenarios [157,158]. As a matter of fact, there are cases in which symbol identity is not enough to measure the similarity between two strings. As an example, let us consider two strings generated by the discretization of two time series arising from heterogeneous data streams, such as two sensors measuring different phenomena, such as light and temperature. Indeed, the values, the scales, and also the meaning of the two sensors may be different but, if these two sensors are near to a fire, light can be influenced by temperature and vice versa. To properly monitor such an event, we should be able to understand the correlations between these two heterogeneous measurements. However, classical string similarity method, based on the assumption of symbol identity, do not take into account such a situation. Thus, in the recent literature, several methods have been proposed to tackle such scenarios [159]. String alignment is a crucial task even in bioinformatics since nucleotide sequences (DNA or RNA) and amino acid sequences (proteins) are encoded as characters. Thus, sequence alignment is the key to understanding the molecular phylogeny of unknown sequences. Molecular phylogeny is accomplished by aligning the unknown sequence with one or more known database sequences to predict the common portions. Alignment is also required for gene annotation by analyzing short RNA-sequence reads derived from messenger RNA (mRNA) and mapping them to the reference genome.

In the following, we present one of the most common and used string similarity metric, the so-called *edit distance*, which is based on the natural assumption of symbol identity. Then, we discuss one of the most recent methods introduced in the literature to address the heterogeneous string-similarity problem, which is called *multi-parameterized edit distance*.

8.2.1 Edit distance

One of the most common string-similarity methods is the so-called edit distance [160]. In 1965 Vladimir Levenshtein introduced a metric for codes capable of correcting deletions, insertions, and reversal, later called *Levenshtein distance* and informally defined as the minimum number of single-symbol edit operations, which are the deletions, insertions, and substitutions required to change one string into another, with each operation having a cost assigned to it. This distance, also often called the metric, belongs to a larger family of distances called edit distance: In such a categorization, different definitions of edit distance use different sets of symbol or string operations. In this section we illustrate the most common setting of edit distance, and we refer to it simply as edit distance.

Given two strings s and p, the edit distance is defined from operations, called edit operations, that transform s into p. Three basic edit operations are considered, namely:

- *insertion* of a symbol of p in s at a given index,
- *deletion* of a symbol of p in s at a given index,
- *substitution* for a symbol of s at a given index by a symbol of p,

A cost, represented by a positive integer value, is usually associated with each of these operations, and the total cost is obtained by adding each of the single edit costs. The edit distance is then defined as the sequence of edit operations to transform s into p that minimizes the total cost of the used operations.

More formally, let s and p be two strings, and let $\Sigma = \text{alph}(s) \bigcup \text{alph}(p)$. For each pair of symbols $x, y \in \Sigma$, we denote the edit operations and their associated cost as follows:

- $f_{\text{ins}}(x)$ is the cost of inserting the symbol x,
- $f_{\text{del}}(x)$ is the cost of deleting the symbol x,
- $f_{\text{sub}}(x, y)$ is the cost of substituting the symbol y for the symbol x.

Having a cost for each edit operation, we denote a sequence of edit operations transforming s to p as α, and the total cost of the edit operations in it as $e(\alpha)$. Finally, let A be the set of possible sequences of edit operations transforming s to p. Then, the edit distance between s and p is given by $edit(s, p) = \min_{\alpha \in A} e(\alpha)$, that is the minimum cost of transforming the string s into p by applying a sequence of edit operations α.

Table 8.1 depicts an example of the edit distance between two strings $s =$ `AGGCTGA` and $p =$ `ACCTG`, along with the edit operations needed to compute it, and their associated costs. In this case, we have $f_{\text{ins}}(x) = f_{\text{del}}(x) = f_{\text{sub}}(x, y) = 1$. The symbol's identity is reflected in the substitution of a symbol x by the same symbol: Indeed, if two symbols are the same, the resulting cost of the substitution is zero, i.e., $f_{sub}(x, x) = 0$. Table 8.1 shows the sequence of edit operations applied to transform s into p, then obtaining $edit(s, p) = 3$.

In the next sections, a recursive formulation of the edit-distance problem is presented and discussed. Then, a classical algorithm for sequence alignment implementing such a formulation, which exploits dynamic programming, is described.

8.2.2 Recursive formulation

With the definition of edit distance, we can now present an algorithm to compute it. In particular, a recursive formulation of the problem can be defined [160]. Let s and

Table 8.1 Example of computation of the edit distance between strings $s =$ AGGCTGA and $p =$ ACCTG. The edit distance is $edit(s, p) = 3$ because the sequence of edit operations necessary to transform s in p cannot be obtained with fewer than *three* edits.

Edit operation	Obtained string	Cost
no operation	AGGCTGA	0
substitution	ACGCTGA	1
deletion	ACCTGA	1
no operation	ACCTGA	0
no operation	ACCTGA	0
no operation	ACCTGA	0
deletion	ACCTG	1

p be two strings with $m = |s|$ and $n = |p|$, and let $d_{m\times n}$ denoting the edit distance between s and p up to indexes m and n, respectively. Then, the edit distance can be defined by the following recurrence:

$$
\begin{aligned}
d_{i0} &= \sum_{k=1}^{i} f_{\text{del}}(s[k]) \quad \text{for } 1 \le i \le m \\
d_{0j} &= \sum_{k=1}^{j} f_{\text{ins}}(p[k]) \quad \text{for } 1 \le j \le n \\
d_{ij} &= \begin{cases} d_{i-1,j-1} & \text{if } s[i] = p[j] \\ \min \begin{cases} d_{i-1,j} + f_{\text{del}}(s[i]) \\ d_{i,j-1} + f_{\text{ins}}(p[j]) \\ d_{i-1,j-1} + f_{\text{sub}}(s[i], p[j]) \end{cases} & \text{otherwise} \end{cases} \\
&\text{for } 1 \le i \le m \text{ and } 1 \le j \le n
\end{aligned}
\tag{8.1}
$$

The recurrence in Eq. (8.1) represents the common recursive way to compute the edit distance between two strings s and p. In particular, the recurrence consists of three main parts. The first one, d_{i0}, computes the cost of transforming s into p by deleting all the symbols $s[1]s[2]\ldots s[m]$, considering the case of the empty string ϵ. Analogously, the second part d_{0j} computes the cost of transforming s into p by inserting all the symbols $p[1]p[2]\ldots p[n]$. The third and last part d_{ij} computes the edit distance between s and p up to indexes i and j, respectively. The recurrence has two cases: The former refers to $s[i] = p[j]$, meaning that the symbols are the same, thus no edit operation is needed and the cost is not modified; the latter case analyzes the cost of each edit operation and selects the minimum obtained one. Finally, the edit distance between s and p is $edit(s, p) = d_{m\times n}$, that is, the edit distance computed up to indexes m and j.

8.2.3 Calculating sequences similarity scores

The computed similarity score provides a measure of how good an alignment is. Alignment of two sequences s and p, where $|s| = m$ and $|p| = n$, produces a new pair of sequences s' and p', such that $|s'| = |p'| = t \mid t \geq m \wedge t \geq n$. The new sequences s' and p' are obtained from s and p by adding a set of "–" dash characters. Similarity defines the resemblance between two sequences, a relevant feature. Although there are many possible ways of arranging letters of two strings (e.g., sequences) to get an alignment, we almost always want the best result: the one with the highest score. The most straightforward approach to calculate a score for two aligned sequences is to estimate how many residue pairs are identical. In this manner, we can measure the sequence similarity.

When calculating how similar two residues are when aligned as a pair, the general idea is to consider how substitutable one residue type is for another, in other words, how likely they are to have been swapped or exchanged for one another. Residues that are commonly interchanged for one another are deemed similar and provide high scores, while those that are rarely switched are dissimilar and give low scores. The similarity score is computed by using a matrix representation, where generally amino acid residues are represented as rows, so that aligned residues appear in successive columns. In an alignment a mismatch can be interpreted as a mutation point, and gaps as insertion or deletion (indels) introduced in one or both sequences to indicate a divergence from one to another.

8.3 Dynamic programming algorithm for edit distances

Various implementations of the recurrence expressed in Eq. (8.1) have been presented in the literature [161,162], starting from 1968. These implementations make use of a dynamic programming [160] algorithm for computing the edit distance, based on the aforementioned recurrence.

Dynamic programming is an algorithmic technique utilized typically in sequence analysis. Dynamic programming efficiently allows solving problems iteratively by storing intermediate results in a table for later use; otherwise, intermediate results are computed repeatedly, which is the primary cause of inefficiency in recursive algorithms. Dynamic programming constructs the n-dimensional equivalent of the sequence matrix formed from two sequences, where n is the number of sequences in the query. Standard dynamic programming is first used on all pairs of query sequences. Then the "alignment space" is filled in by considering possible matches or gaps at the intermediate positions, eventually constructing an alignment essentially between each two-sequence alignment. Dynamic programming is computationally expensive due to the computation of the global optimum solution.

Algorithm 8.1 illustrates the dynamic programming algorithm based on that of [161] that computes the edit distance between two given strings s and p. The main idea at the core of such algorithm consists of a matrix employed to reserve the computed edit distance between all prefixes of the two strings. Having these values, the rest of the matrix can be traversed by a common flood-filling algorithm, which computes the distance following the cases depicted in the third part of the recurrence in Eq. (8.1). We can now briefly detail the algorithm and its behavior.

Algorithm 8.1: Dynamic programming algorithm to compute the edit distance between two given strings s and p.

Input : Strings s and p of length m and n (with indices $[1..m]$ and $[1..n]$ resp.); w_{ins}, w_{del}, w_{sub} the cost for insertion, deletion, and substitution edit operations, respectively

Output: Edit distance between s and p

```
d ← [0..m, 0..n] matrix
for i ← 0 to m do
    for j ← 0 to n do
        d[i, j] ← 0
for i ← 1 to m do
    d[i, 0] ← i
for j ← 1 to n do
    d[0, j] ← j
for j ← 1 to n do
    for i ← 1 to m do
        if s[i] = p[j] then
            w ← 0
        else
            w ← w_sub
        d[i, j] ← min(
                d[i − 1, j] + w_del,
                d[i, j − 1] + w_ins,
                d[i − 1, j − 1] + w
        )
return d[m, n]
```

Algorithm 8.1 receives as input two strings s and p of length m and n, respectively, and the cost for each edit operation, i.e., w_{ins}, w_{del} and w_{sub}, respectively, for insertion, deletion, and substitution. Line 1 defines a $m + 1 \times n + 1$ matrix, whose purpose is to store the edit distance between all prefixes of the strings in input. The first fors at line 2–3 fill the matrix with the value 0 in each cell. Then, the for at line 5 denotes the cost of transforming the prefixes of s into ϵ by deleting every symbol in it. Similarly, line 7 depicts the code for computing the cost of transforming ϵ into p by inserting every symbol in it. The main part of the recurrence in Eq. (8.1) is then implemented by the lines 9–19. In particular, the algorithm iterates over the matrix and, for each cell, the symbols in the respective positions in s and p are checked. If $s[i] = p[j]$, then they are the same symbol, and therefore there is no cost for substituting them. The cell $[i, j]$ in the matrix indicates the edit distance between s and p up to indices i and j, respectively, which is computed by the recurrence. Finally, the bottom-most cell on the right at indexes $[m, n]$ contains the edit distance between s and p, which is then returned as output of Algorithm 8.1.

The time complexity for the implementation given in Algorithm 8.1 is $\mathcal{O}(mn)$. Several pragmatic variants have been proposed in the literature that achieve slightly lower time complexity. As an example, if we are interested in the distance only in the case that it is smaller than a threshold θ, it is thus possible to employ an algorithm with time complexity $\mathcal{O}(\theta l)$, where $l = \min(|s|, |p|)$, with s and p being the input strings [163].

8.3.1 Needleman and Wunsch algorithm

The optimal alignment problem can be formulated as the alignment formed by the smallest possible number of differences necessary to transform one string into another, or more generally considering the shortest path separating a string from the other. Calculating the optimal alignment by means of an exhaustive procedure is impractical because the number of combinations grows exponentially, even for small sequences. For this reason, in [162], Needleman and Wunsch (N & W) have proposed a dynamic method for solving this problem.

In the method proposed by N & W, the atomic comparison unit is the symbol-pair. The best method to represent all the possible pairing is using a matrix, where the two sequences are arranged on both sides of the matrix, and each symbol identifies a row or column cell, as depicted in Fig. 8.2.

Otherwise, the cell identified by the same character is initialized to 1, 0. A possible alignment between two sequences in this manner is determined by the path within the matrix joining the cells corresponding to the paired symbols (see Fig. 8.3).

Because we want to align biological sequences with specific characteristics such as polarity, the path must satisfy some requirements. It must proceed forward, diagonally towards the C-Terminal in the case of proteins, vertically or horizontally for the indels, without ever going back (see Fig. 8.3). Furthermore, each amino acid can occupy only one position and be touched by the path only once. Given the following constraints, the problem of seeking and finding the best alignment between two sequences can be formulated as follows. It is necessary to identify, among all the possible paths within the matrix, the one containing the most significant possible number of identical amino acid residues with the least likely number of indels. To this end, we can use the following scheme, where the value 0 represents two paired residues, 1 indicates a residue in the first sequence and a gap in the second sequence, and 2 vice-versa a gap in the first sequence and a residue in the second sequence (see Fig. 8.2). This convention allows the algorithm to determine whether to assign a penalty, just by checking if the previous column or row contains a gap (coded as 1 or 2). The Python function of N & W algorithm is shown in Appendix A.

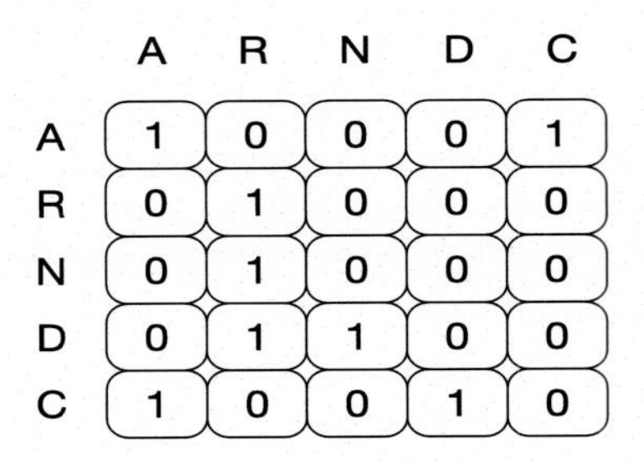

FIGURE 8.2

A simple matrix initialized for the alignment.

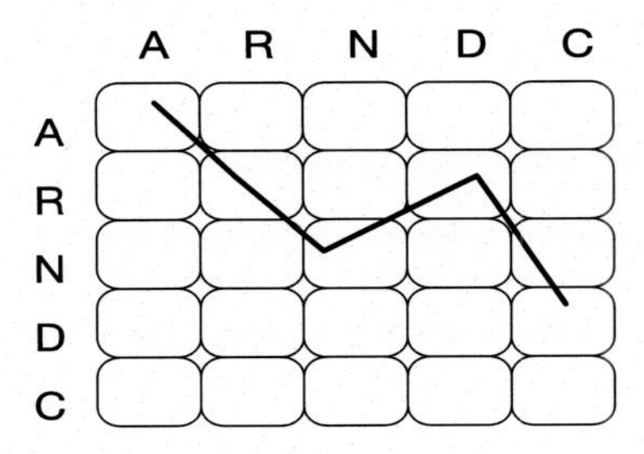

FIGURE 8.3

A possible path into the matrix, highlighting the paired residues.

8.4 Multi-parameterized edit distances

The edit distance, as well as other symbol-identity based string similarity methods, can be applied perfectly in contexts preserving the assumption that identical symbols represent identical information. However, as we already pointed out, this assumption is not valid in several contexts.

Let us look at an example. Let $s = \texttt{AAABCD}$ and $p = \texttt{444567}$ be two strings, and let the edit distance be our method of reference. According to it, s and p are two completely different strings, e.g., the distance between them is the maximum possible, that is $edit(s, p) = 6$, considering the cost of 1 for each edit operation. Nevertheless, carefully looking at the strings seems to yield some insight into their similarity. As an example, if we define the symbols A and 4 as "identical", i.e., we assume they represent the same information, then the edit distance clearly differs, returning a value of 3. Therefore, in the context of heterogeneous strings, it is possible to adapt the edit distance, as well as other string similarity methods, by providing a sort of mapping between the symbols. Such a mapping could state which symbols match and thus be considered as identical, e.g., representing the same information.

The Multi-Parameterized Edit Distance (MPED) is a novel string-similarity method based on two main elements: the classical edit distance method and the concept of matching schema. A matching schema resembles the idea of mapping we just introduced: It represents a mapping between two (possibly disjoint) alphabets and

$M_1 = \{$(C,K), (A,Z), (D,4), (1,3), (0,Y)$\}$ ⟶

```
---CCADD110
ZZZKK--4Y3-
   **   * *
```

$M_2 = \{$(C,Z), (A,4), (D,K), (1,Y), (0,3)$\}$ ⟶

```
CCADD110
ZZZKK4Y3
**  **  **
```

FIGURE 8.4

Example of matching schema with their corresponding edit distances.

states which symbols of the first alphabet can be considered matching with symbols of the second alphabet. Such a mapping makes it possible to identify hidden correlations between strings and can be used to define a similarity between them. Indeed, the matching schema is used within the computation of the edit distance, where the question "are two symbols the same?" becomes "are two symbols matching w.r.t. the matching schema?".

To better understand the MPED, in the following sections we present the background behind the MPED and an algorithm to compute it. Due to a lack of space, here we only briefly discuss it. We refer the interested reader to [157,164] for a more detailed discussion and an overview of applications.

8.4.1 Background

Having introduced the core idea behind the MPED, let us now briefly provide a formal background to it. Let Σ_1 and Σ_2 be two alphabets, and let s and p two strings defined on Σ_1 and Σ_2, respectively. A matching schema M between Σ_1 and Σ_2 states the mapping between the symbols in Σ_1 and the symbols in Σ_2. We here consider this mapping as a bijection between the two alphabets, thus $|\Sigma_1| = |\Sigma_2|$. However, several applications require different mappings, and we refer the interested reader to [157] for a comprehensive overview. We represent M as a set of pairs (a, b), $a \in \Sigma_1$, $b \in \Sigma_2$, where each pair denotes that the symbol a from Σ_1 matches with the symbol b from Σ_2, e.g., they are considered identical. We denote as $\mathbb{M}_{\Sigma_1\Sigma_2}$ the set of the possible matching schemas that can be built over the two alphabets, i.e., all of the possible bijections between Σ_1 and Σ_2. We simply write $\mathbb{M}$ when it is clear from the context.

A matching schema M is then employed in the computation of the edit distance. In particular, the condition of symbol identity, e.g., line 11 in Algorithm 8.1, is then replaced by the mapping dictated by M. We denote computation of the edit distance between two strings s and p, enhanced with a matching schema M, as $edit^M(s, p)$. Then, we define the MPED between s and p as the minimum edit distance that can be obtained by any $M \in \mathbb{M}$, i.e., $MPED(s, p) = \min_{M\in\mathbb{M}}\{edit^M(s, p)\}$.

Before presenting an algorithm for the computation of the MPED, let us provide an example. Let $s =$ CCADD110 and $p =$ ZZZKK4Y3, respectively, defined on $\Sigma_1 = \{$0, 1, A, C, D$\}$ and $\Sigma_2 = \{$3, 4, K, Y, Z$\}$. If we take into account the classical edit distance, we obtain $edit(s, p) = 8$, which results from substituting every symbol of s. Now, let us consider two matching schemas $M_1 = \{$(C, K), (A, Z), (D, 4), (1, 3), (0, Y)$\}$

and $M_2 = \{(\texttt{C}, \texttt{Z}), (\texttt{A}, \texttt{4}), (\texttt{D}, \texttt{K}), (\texttt{1}, \texttt{Y}), (\texttt{0}, \texttt{3})\}$, depicted in Fig. 8.4. A tuple in a matching schema indicates that two symbols match. For instance, the tuple $(\texttt{C}, \texttt{K}) \in M_1$ states that the symbol C from Σ_1 matches with the symbol K from Σ_2. Then, if we employ these matching schemas in the computation of the edit distance, we obtain $edit^{M_1}(s, p) = 7$ and $edit^{M_2}(s, p) = 2$, respectively. Fig. 8.4 shows the obtained edit distance taking into account the matching schema, where the colors group together their stated mappings.

8.4.2 An algorithm for computing the MPED

In the example in Section 8.4.1, it is straightforward to observe that the matching schema giving the lowest edit distance is M_2. It is important to point out that, in the general case, computing the MPED between two strings s and p is a NP-Hard problem [157]. Indeed, several methods exploiting metaheuristic approaches have been presented in the literature to tackle the hardness of the problem [157,158,164].

In this section, we illustrate an algorithm employing a metaheuristic presented in [164]. The algorithm is based on a very well-known metaheuristic called Simulated Annealing [165] (SA, from now on). It represents a generic probabilistic metaheuristic for optimization problems presenting a very large search space. The goal of SA is to bring a system, from an initial state, to a state with the minimum possible energy. In this approach, given two strings s and p and their corresponding alphabets Σ_1 and Σ_2, a state of the SA is a matching schema in $\mathbb{M}(\Sigma_1, \Sigma_2)$, and the state corresponding to the minimum possible energy is the matching schema giving the lowest edit distance between s and p.

Algorithm 8.2: Pseudocode of the simulated annealing metaheuristic for computing the MPED between two strings.

Input : Strings s and p; alphabets Σ_1 and Σ_2; temperature parameter t
Output: MPED between s and p

```
M ← random(𝕄(Σ1, Σ2))
curr ← edit^M(s, p)
while t > 1 do
    t ← t − 1
    M' ← generate(M)
    new ← edit^M'(s, p)
    if curr > new or prob() then
        M ← M'
        curr ← new
        if best > curr then
            best ← curr
return best
```

Algorithm 8.2 shows the pseudocode of SA. It takes as input the two strings and the corresponding alphabets. Moreover, it also takes as input a parameter t called temperature. This parameter can be decided by the user. Then, SA proceeds as follows. It starts by selecting a random matching schema M from $\mathbb{M}(\Sigma_1, \Sigma_2)$, and computes the edit distance, according to M, between s and p. Then, until the temperature parameter has a value greater than 1, the following steps are made. First, a matching schema M^r is generated by performing a random perturbation of the current matching schema M. Then, the edit distance according to M^r is computed. The condition at line X holds if the computed edit distance is lower than the current one; otherwise, a probability function based on the energy of the current and the next states is used [165]. If the condition holds, then M becomes M^r, and the best distance is updated as well. Finally, the SA returns the best edit distance between s and p, e.g., the lowest one.

8.5 Alignment free sequence comparison

Alignment-free (AF) [166] methods have emerged as an alternative to alignment-based approaches for quantifying similarities between sequences. Alignment-based approaches are computationally expensive [167], especially in analyzing next-generation (NGS) sequences because they are time-consuming and depend on knowledge of the whole genome. NGS sequences may not be fully sequenced due to the stochastic distribution of the reads along the genomes and the difficulties of sequencing some regions, especially when the coverage is relatively low. Thus, these lacks pose significant challenges for alignment-based approaches due to the considerable data size and the relatively short length of the reads, as reported in [168]. As a result of these limitations, alignment-free approaches are emerging as an alternative to alignment-based methodologies. Alignment-free approaches based on the words-count method are independent of the whole sequence, making them computationally efficient and suitable for aligning NGS data.

The word-counts [169,170] method is a two-step procedure. The first step counts the number of occurrences of word patterns (k-mers, k-grams, k-tuples) within the NGS sequence. Second, a measure of similarity or dissimilarity computes the alignment on the pattern frequencies between any pair of sequences and puts similar sequences in the same group using a clustering algorithm. A well-known word-count algorithm is DSK (disk streaming of k-mers) [171], an efficient and effective word-count method for NGS sequences.

The available k-mer counting methods need to store k-mers count in a massive hash table in the main memory. Hence, the hash table grows with the distinct k-mers requiring counting. To overcome this limitation, the DSK (disk streaming of k-mers) method defines a trade-off between the amount of used memory, disk space, and time. The DSK approach realizes a memory, time, and disk trade-off. DSK partitions the multi-set of all k-mers present in the reads, saving the multi-sets to a disk. Then, each partition is separately loaded in memory in a temporary hash table only when necessary. Algorithm 8.3 shows the main steps of the DSK method.

Algorithm 8.3: DSK pseudo code.

Require: The sequences $\varsigma = \{S_1, \ldots, S_n\}$, k-mer length, M_{size} percentage of memory to use (MB), the disk space (MB), the hash function $h(\cdot)$
Ensure: LA List of alignments
1: $n_{kmer} = computeKmerNumber()$
2: $n_{iter} = defineIterationNumber()$
3: $n_{part} = definePartNumber()$
4: $j = 0$
5: **for all** $i \in n_{iter}$ **do**
6: $\kappa = \{d_1, \ldots, d_{n_{part}}$ {generate a kmer list on disk}
7: **for all** $s \in \varsigma$ **do**
8: **for all** $k \in n_{kmer}$ **do**
9: **if** $h(k)\%n_{iter} = i$ **then**
10: $j = \frac{h(k)}{n_{iter}}\%n_{part}$
11: **end if**
12: $write(k, d_j)$
13: **for all** $j \in n_{part}$ **do**
14: $\tau = createHashTable(M_{size})$ {creates initializes a hash table using M_{size} memory bits}
15: **end for**
16: **for all** $km \in \varsigma$ **do**
17: **if** $km \in \tau$ **then**
18: $\tau[km]++$
19: **else**
20: $\tau[km] = 1$
21: **end if**
22: **end for**
23: **end for**
24: **end for**
25: **end for**
26: **end.**

CHAPTER

Structure analysis

9

CONTENTS

Abstract

Proteins and nucleic acids have a spatial structure whose analysis is a central topic in bioinformatics. Such analysis has two major goals: predicting the structure of single molecules and simulating the interaction among two or more molecules. The first methods for structure prediction were based on statistical and mathematical approaches. Nowadays, machine and deep learning approaches have been developed, while more recent artificial intelligence approaches appear to represent a breakthrough. We here define the problem, and we discuss some of the main predictive methods for one-dimensional and two-dimensional structure prediction. We conclude this chapter by discussing the challenges and opportunities that may arise next.

Keywords

Primary structure, Secondary structure, Tertiary structure, Quaternary structure, Secondary structure prediction, Tertiary structure prediction

9.1 Introduction

The spatial structure of a protein is closely related to its function. Consequently, the study of such structures has great importance.

Protein structures are usually organized into four levels: primary, secondary, tertiary, and quaternary, as depicted in Fig. 9.1.

Artificial Intelligence in Bioinformatics. https://doi.org/10.1016/B978-0-12-822952-1.00019-X

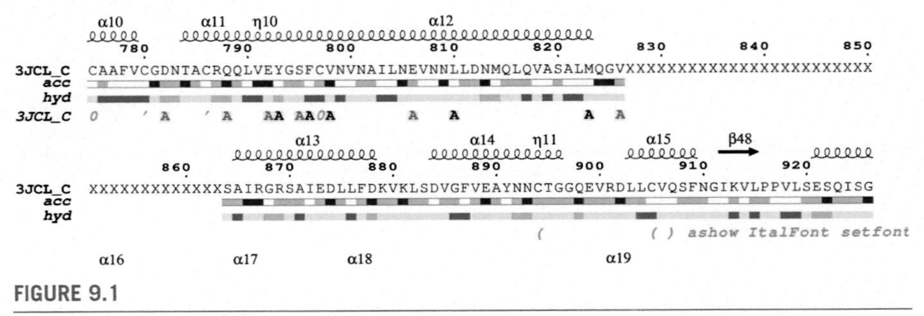

FIGURE 9.1

Representation of the secondary structure of the human 3JCL depicted by using the ENDScript Web Server [172]. The secondary structure is superimposed on the primary one.

The linear sequence (also referred to as a polypeptide chain) of amino acids that constitute a protein is the **protein primary structure**. The local structure of the polypeptide chain is the **protein secondary structure (PSS)**, and it is generated by the action of hydrogen bonds among peptides [173].

On the basis of the characteristics of the polypeptide chain, there exist many different secondary structures. Initially, computational biologists shed light on three main structures: Helix (H), Strand (E), and Coil (C). Helix is determined by the helical configuration of peptides caused by the formation of a hydrogen bond between every fourth amino acid. Strand is a segments structure of linked parallel or antiparallel caused by bonds among interacting amino acids. Finally, C is a class comprising all the amino acids that do not belong to previous classes [174].

Subsequently, the DSSP (Dictionary of Secondary Structure of Proteins) [175], introduced a novel categorization based on eight structures: H (α-helix), G (310-helix), I (π-helix), E (β-strand), B (isolated β-bridge), T (turn), S (bend), and C (others).

The **protein tertiary structure** refers to the spatial configuration obtained by the fold of multiple secondary structures in 3-D space. The protein folding process represents the way to perform all the biological functions, and it aims to minimize the overall free energy. The folding is completely determined by the primary and secondary structures. The quaternary structure refers to protein complexes, i.e., the binding of two or more proteins, and it represents the characteristics of its biological function.

9.2 Protein secondary structure prediction

The determination of protein structure may be done through in-vivo experiments, such as X-ray crystallography or nuclear magnetic resonance. Because these processes require a considerable amount of time and resources, the possibility to predict such structures is a key challenge in computational biology.

Protein secondary structure prediction (PSSP) refers to the set of methods, algorithms, and implementation tools that are used to predict the secondary structure of a protein starting from the polypeptide chain.

Prediction methods can be categorized in three main classes: (i) ab initio calculus of minimal energetic conformation, (ii) statistical elaboration over known structures, and (iii) employment of neural networks.

The first group is based on the theoretical consideration that the native folding is an energetic minimum. It has been demonstrated that the determination of the energy minimum is a NP-complete problem. For this reason, algorithms belonging to this class tighten the research space with biological considerations.

An example of second group is the method of Chou and Fasman. The idea is that each amino acid has a particular tendency to participate in a particular secondary structure (that is, to a random coil). It is given by the relationship among the fractions of residuals that are found in that secondary structure and the fraction of residues that is found in each of the three structures (helix, sheet, coil).

The **GOR** method, developed by Garnier et al. [176], is based on the consideration that the secondary conformation is the result of an equilibrium. Each amino acid is influenced by nearest neighbors, and the tendency to participate in a particular structure is modified.

9.2.1 Artificial neural networks for structure prediction

Neural networks used for PSSP receive as input the primary structure of proteins (and other information in some cases) and predict the secondary structure, as depicted in Fig. 9.2.

Neural networks are used for structure prediction to automatically build the relationship between the primary and secondary structures [177]. Generally, the application of neural networks is a mapping between an input vector representing protein primary structure and the secondary structure centered around a central peptide in a small window.

Qian and Sejnowski proposed one of the earliest methods based on neural networks for PSSP in 1988 [178]. For many years, the standard feedforward backpropagation networks have been primarily used in this domain. The general function learned by neural networks has the following representation:

$$s(x) = f \sum_{i=1}^{n} \omega_i m_i + q_i, \tag{9.1}$$

where $s(x)$ is the output of the prediction (i.e., a predicted structure that may have three or eight values), f is the activation function, m_i is the activate value, ω_i is the associated weight, and q_i is the intercept. Usually, the sigmoid function or the tanh is

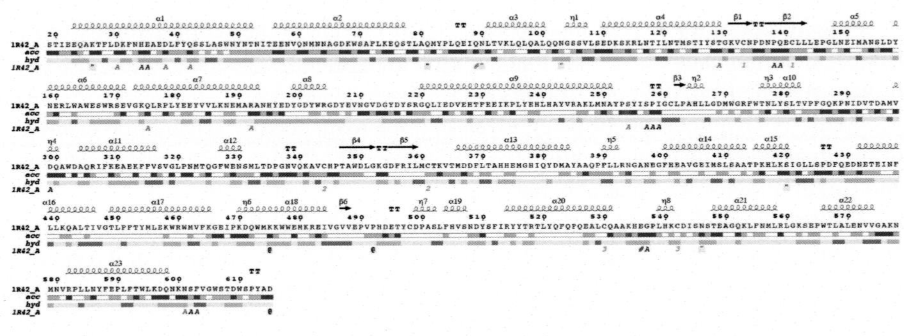

FIGURE 9.2

Figure represents the superimposition of the primary structure of the ACE-2 human protein and its secondary structure, as depicted by ENDScript Web server [172]. PSSP software receives as input the string representing the primary structure and produces as input the secondary structure fragments.

used. Consequently, m_i may be written as:

$$m_i^{l+1} = f\Big(\sum_{j=i}^{n} \omega_j^l x_j + q_j^l\Big), \tag{9.2}$$

where l is the l-th hidden layer and j represents the j-th node of the l-th hidden layer that is connected to the i-th node of the $l+1$ hidden layer.

PHD (Profile from Heidelberg) method [179] receives as input a set of aligned sequences and not a single sequence. It is based on three layers: (i) sequence to structure, (ii) structure to structure, and (iii) jury decision. The first layer is a neural network that classifies the central residue of the entire multiple alignment. The second formalizes the consideration that the input sequences are correlated. The last layer improves the performances by combining results from 12 different neural networks.

JPRED is a consensus method for protein secondary structure prediction [180]. It is freely available and is implemented on a web server. It uses a neural network called Jnet. It receives as input a multiple alignment or a single sequence. In this case, a multiple alignment is preliminarily calculated.

Actually, there exist many network architectures that have shown good performance in predicting secondary structure: deep learning, recurrent neural networks (RNN), complex-valued neural networks, BP neural network, and radial basis function neural network [177].

For instance, bidirectional RNNs (BRNNs) have been used in Porter and PaleAle 4.0 by Mirabello and Pollastri [181]. In these softwares two cascaded neural networks are used. The first BRNN is used to predict the secondary structures from the primary sequence and multiple sequence alignments. The second BRNN is used to filter and improve the prediction obtained by the first BRNN. 2D recursive neural networks are used in [182]. In this work protein inter-residue contact maps are used. Another successful RNN based method is SPIDER3 [183], which is an improvement on previous work. SPIDER3 was based on a BRNN model that contained long short-term memory (LSTM) cells. LSTM cells improve the prediction since they are able to include in the prediction both local and non-local relationships among fragments of the sequence.

Each residual block is based on a 3 x 3 convolutional neural network. The combination of all the inputs gives to the model more than 21 millions parameters.

9.3 Tertiary structure prediction

Tertiary structure is the general spatial conformation, while the quaternary one is the structure determined from more polypeptidic chains. Knowing protein structures is useful for studying protein functions and their roles in chemical reactions.

9.3.1 Databases of structural classifications

The structural domain of a protein is an element of the ternary structure that often folds independently of the rest of the protein chain and constitutes a biological relevant module of the proteins themselves. Despite the existence of a very large number of proteins expressed in eukaryotic systems, there are much fewer different domains, structural motifs, and folds. Many domains are not unique to the proteins produced by one gene or one gene family, but instead they appear in a variety of proteins as a consequence of evolution that has conserved spatial conformation better than the primary sequence.

Consequently, several methods have been developed for the structural classification of proteins, and a number of different databases has been introduced, such as CATH [184] and SCOP [185].

9.3.1.1 SCOP

The Structural Classification of Proteins (SCOP)[1] database aims to order all the proteins whose structure has been published according to their structural domains. Protein domains in SCOP are hierarchically classified into *families, superfamilies, folds, and classes*. First, proteins are grouped together into families on the basis of evolutionary or functional similarities, such as sequence alignment. Then, proteins whose sequences have low similarities, but whose structure or functions are close, are grouped into superfamilies. The secondary structure of the proteins in these two groups is analyzed, and when the proteins in two groups, e.g., two families, have similar secondary structure, they have a common fold. Finally, folds are grouped into classes: (i) All α if the structure is essentially formed by α-helices, (ii) all β

[1] http://scop.mrc-lmb.cam.ac.uk/scop.

if the structure is essentially formed by b-sheets, (iii) a/b for those with α-helices and β-strands, (iv) a+b for those in which α-helices and β-strands are largely segregated, and (v) multi-domain for those with domains of different classes. The SCOP database is available on the World Wide Web (WWW) where a user has many options to browse its content. The main possibility is to access from the top of the hierarchy and then navigate through the levels, from root to the leaves that are individual PDB entries. Alternatively, the user can search a protein starting from an amino acid sequence to retrieve the most similar proteins categorized in SCOP. The user can download the found protein as a single PDB file.

9.3.1.2 CATH

CATH[2] [184] stores a hierarchical classification of PDB structures obtained with NMR and crystal structures resolved to resolutions higher than 4.0 angstroms. Protein structures are classified using a combination of automated and manual procedures on the protein domains. To divide multidomain protein structures into their constituent domains, a combination of automatic and manual techniques is used. If a given protein chain has a sufficiently high sequence identity and structural similarity, the hierarchy is organized in four major levels: *Class, Architecture, Topology (fold family), and Homologous* superfamily. Class is determined according to the secondary structure composition, and currently three major classes have been recognized: mainly-alpha, mainly-beta, and alpha-beta that groups both alternating alpha/beta structures and alpha+beta structures. A fourth class contains protein domains that have a low secondary structure content. The architecture level considers the overall shape of the domain structure as determined by the orientations of the secondary structures, and it is assigned manually. The topology level groups the structures depending on the overall shape and on the connectivity of the secondary structure applying the algorithm SSAP [186]. Finally, the Homologous Superfamily level groups protein domains that are thought to share a common ancestor and can therefore be described as homologous as recognized by SSAP. Currently, it contains 30,028 PDB structures. CATH contains PDB structures organized in a relational model. CATH can be searched by submitting a protein identifier, or by browsing the hierarchical structure. Moreover, users can access data via FTP and can also download them.

9.3.2 Algorithms for tertiary structure analysis

Algorithms studying such data can be grouped in: (i) **Alignment Algorithms** that compare two or more spatial structure, (ii) **Fold Prediction Algorithms** that predict the folding of proteins, and (iii) **Docking Algorithms** that predict the possible binding of two proteins on the basis of their structure. Fig. 9.3 depicts a taxonomy of these algorithms.

Protein-structure alignment techniques have grown increasingly important as a means to quantitatively compare and classify all known protein structures. One of

[2] http://www.cathdb.info/latest/index.html.

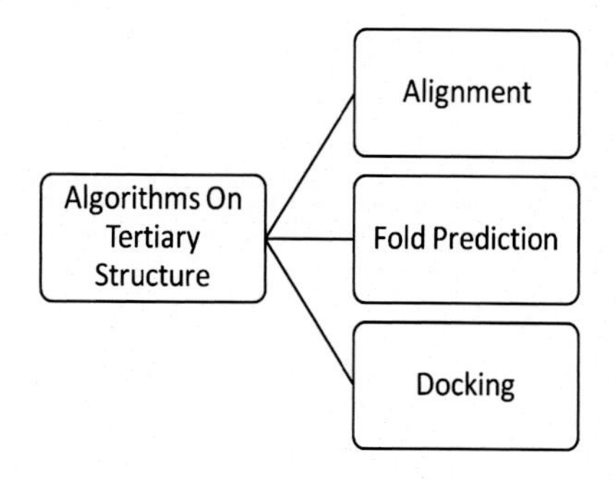

FIGURE 9.3
Taxonomy of algorithms.

the primary goals of structural alignment programs is to quantitatively measure the level of structural similarity between all pairs of known protein structures. This data can provide several meaningful insights into the nature of protein structures and their functional mechanisms.

For instance, the comparison of all structures against each other can show relationships, both functional and structural, between proteins that were previously not known to be related. In addition, structure-based distance measures are critical to constructing accurate phylogenies of proteins and classifying structures into families that share similar folds or motifs.

There have been several methods proposed to compare protein structures and measure the degree of structural similarity between them. These methods have been based on alignment of secondary structure elements, as well as alignment of intra- and intermolecular atomic distances. In the second case the problem can be formulated as a geometric problem in terms of Root-Mean-Square Deviation (RMSD) measure between the amino acids of the back bone, represented as points in a three-dimensional space. Given two sets P and Q, each one containing 3-D points, we need to find a rotation P and a translation Q to minimize the RMSD of distances between corresponding residues. This problem has been solved with dynamic programming algorithms.

An example is **SSAP** [186] developed by Taylor and Orengo in a technique called double dynamic programming. The steps are the following: (i) define a local invariant structural environment for each residue; (ii) compute the similarity/distance for each pair of residues; (iii) consider each computed distance as an entry in the dynamic programming matrix; (iv) find the optimal path in the matrix.

Flex-Prot [187] algorithm receives two protein molecules A and B, each one represented by the sequence of the 3-D coordinates of its carbon atoms. It finds the largest flexible alignment by decomposing the two molecules into a minimal number of rigid fragment pairs having similar 3-D structure. An alternative approach is represented by the techniques of geometric hashing [188], based on transformation invariant representations. The DALI (Distance mAtrix aLIgnment) [189] approach receives the coordinates of a protein in PDB (Protein Data Bank) format and compares them with all the proteins stored in a database.

9.3.2.1 AI approaches for tertiary structure prediction

Methods for protein structure prediction fall into two major classes: template-based models and template-free models. The first class of algorithms is based on the use of a template structure that is used as reference for building the prediction. The second class is used when a template is not available or an existing template does not have good accuracy.

Template-based modeling is typically more accurate if a good template can be found because it exploits existing protein structures. Template-free models are subsequently divided in two categories: fragment-based assembly and ab initio or de novo folding. The former is based on the assembly of existing fractions of templates, while the latter tries to reconstruct the 3-D shape by using a simulation approach based on chemical and physical principles. Such methods rely on the simulation of the minimization of the energy function, through an efficient conformational search algorithm.

However, fragment-based assembly is the dominant technique due to its higher accuracy and higher capability whenever a good template is difficult to identify or otherwise unavailable.

AlphaFold [190] is a recently developed method for protein structure prediction that achieved the best performance in the ab initio category of CASP13. The method is based on the prediction of the distance map among residues using a deep residual neural network. It used 220 residual blocks that are equivalent to 64 input amino acids.

AlphaFold's top score in 25 out of 43 test proteins has excited researchers about what the future may hold for AI-based protein structure prediction.

The neural network of the AlphaFold system is able to predict the spatial coordinates of all the atoms of a given protein. It receives as input the primary sequence of the amino acids and a multiple alignment of the homologous ones and a list of templates (a detailed description of the system can be found on https://static-content.springer.com/esm/art10.1038Fs41586-021-03819-2/MediaObjects/41586_2021_3819_MOESM1_ESM.pdf).

The neural network has two main components: The first one processes the sequences and the templates, while the second one transform the sequence into 3-D coordinates and refines the predicted structure. The core idea of these modules is to predict protein structures through the solution of a graph inference problem in 3-D space in which the edges of the graph are defined by the residues in proximity. The elements of the pair representation encode information about the relationships between the residues.

CHAPTER

Omics sciences 10

CONTENTS

Abstract

In the last two decades, the use of words ending in "omics" has extended from the initial genomics to a wide range of biomolecular disciplines, even entering everyday language usage. In general, the suffix "omics" signifies to study specific aspects considered as a whole and/or on a large scale. The spread of omics disciplines has been made possible mainly because of the development of High-Throughput (HT) investigation technologies capable of generating enormous quantities of data relating to the multiple hierarchical levels of biological complexity (DNA, mRNA, proteins, metabolites, etc.), helping to revolutionize the study of living beings. A fundamental challenge to face in analyzing omics data is the vastness and heterogeneity of these data, making the arising problems computationally impractical. Fortunately, robust algorithms and techniques can create software that can handle critical biomedical enigmas in practice. Subsequently, a wide assortment of algorithms and software has been over the past two years. Still, for anyone to extract biological insights from these data sets, experience with increasingly sophisticated computational techniques is required. This chapter describes the main "omics" disciplines, the technologies related to them, and the main application areas.

Artificial Intelligence in Bioinformatics. https://doi.org/10.1016/B978-0-12-822952-1.00020-6

Keywords

Next-generation sequencing, DNA, Genomics, Transcriptomics, Epigenomics, Proteomics, Metabolomics

10.1 Introduction

In the last two decades, the use of words ending in "omics" has extended from the initial genomics to a wide range of biomolecular disciplines, also entering usage in everyday language. In general, the suffix omics signifies the study of specific aspects considered as a whole and or on a large scale. The spread of omics disciplines has been made possible mainly because of the development of High-Throughput (HT) investigation technologies capable of generating enormous quantities of data relating to the multiple hierarchical levels of biological complexity (DNA, mRNA, proteins, metabolites, etc.), helping to revolutionize the study of living beings. Understanding how DNA works can contribute to identifying genes and proteins responsible for diseases and drug interactions. Thus, the development of omics science such as genomics, proteomics, epigenomics, and metabolomics has become crucial for understanding the causes and the treatment of diseases to maintain human health. Significant advances in omics science have been possible, ranging from traditional reductionist approaches based on understanding complex phenomena through their disaggregation to single and more straightforward aspects and toward systemic methods with a holistic vision of such complex phenomena. The leading omics science disciplines comprise the following: genomics that refers to the complete nucleotide sequence, i.e., the genetic makeup of an organism; proteomics that concerns the whole proteins of a cell in any living organism; epigenomics that is the study of the nucleotide changes within an organism; proteomics that comprises to the structural and functional study of the entire collection of proteins expressed in a specific cell type or organism, called a proteome; and metabolomics that investigates the changes in gene activity in response to metabolites. In the following a schematic and concise overview of the main "omics" disciplines, the technologies related to them, and the main application areas are reported.

10.2 Genomics

Genomics [191] is understood as the discipline that deals with the study of the genome in a systematic way. Appearing mainly following the emergence of in vitro DNA amplification techniques by Polymerase Chain Reaction (PCR) [192] and the spread of automation systems for DNA sequencing technologies (e.g., Sanger's method [193]), the new discipline initially had large-scale genetic mapping and sequencing as the primary goal. Over time, the omics term assumed a broader connotation, including structural genomics studies (aimed at predicting the 3-D structure of proteins, starting from information on the nucleotide sequence), the analysis of single nucleotide polymorphisms (SNPs), the study of comparative and evolutionary genomics (based on the comparison between genomes of different species aimed at understanding the evolutionary mechanisms underlying the genetic relationships between organisms), and functional genomics (aimed at understanding the function and interactions between genes and their products of expression through, e.g., the study of transcription, translation, and protein/protein interactions).

From a technological point of view, one of the most relevant innovations in genomics is undoubtedly represented by the extensive development of numerous Next-Generation Sequencing (NGS) approaches. The significant advance offered by NGS is its ability to produce considerable volumes of data cheaply per single experiment, unimaginable with traditional techniques.

10.2.1 Next-generation sequencing

The NGS analysis workflow involves the following steps:

1. The construction of sequencing libraries from DNA or RNA molecules obtained from biological samples of interest. Generally, the extracted DNA or RNA molecules need to be broken into smaller fragments because their molecules are too long to be directly handled by NGS assays.
2. Fragmentation can be achieved by various techniques, e.g., sonication, nebulization, or enzymatic treatment.
3. Fragment collection regards the collection of fragments with a precise length that is the number of bases.
4. RNA fragments conversion. RNA fragments are usually converted to complementary DNA (cDNA) before being added to the adapters. The transformation and subsequent adapter ligation to the two ends of DNA fragments are two essential sub-steps in the process of the sequencing library construction.
5. Amplification is accomplished by using the PCR for sequencing template enrichment.
6. Sequencing. This step concerns the sequencing of the constructed DNA library using NGS technologies. In this manner, it is possible to generate a massive number of short reads varying in length between 40 and 200 base pairs.
7. The reading of short reads using optical or physicochemical signals leads to the sequence deduction of the DNA fragments.
8. Saving of the NGS analysis run results in a file coded in a specific format, to enable their accurate analysis by using high-performance machine learning and data mining algorithms.

The NGS output is usually stored in a platform-specific format. In contrast, the processed sequence reads are typically reported in a universal file format.

FASTQ is the standard format used to store base-calling results. In addition, other NGS file formats are available, e.g., FASTA, CFASTA, SFF, and QUAL. Because FASTQ is the de facto standard for storing base-calling data, the other NGS file formats can be easily converted to FASTQ using conversion tools (such as *NGS QC Toolkit*, for example). The typical size of a compressed FASTQ file is some gigabyte and may contain about a million or more reads in a single file. Concisely, FASTQ files, including the sequence of each read along with the confidence score of each base. A FASTQ file uses four lines to represent a sequence. In the first line, the sequence identifier starts with the '@' character and is followed by a description (optional). The second line contains the raw sequence letters. The third line (optional) contains the sequence identifier and any description and begins with a '+' character. The fourth line encodes the quality values for the sequence in line 2. Fig. 10.1 displays an example of base-call read in the FASTQ file format.

```
@SRS123456.1 30443AAZZ:1:1:1024:1042 length = 20
ACGTGCATACAGTCACGTAG
+
=IIIIIIII2IIIIIIIII>III+AGIIIIIIIIII(IIIIII
```

FIGURE 10.1

Overview of FASTAQ file content.

SAM (Sequence Alignment Map) and BAM (Binary Alignment Map) are other standard formats used to store reads that map results concerning a reference genome. SAM is a tab-delimited text format, human-readable, and easy-to-explore but relatively slow for automatic processing. BAM is the compressed binary version of SAM. BAM is smaller in dimension and faster to process. The basic structure of a SAM/BAM file is straightforward. Both contain a header section (optional) and an alignment section. Before the alignment section, the header section provides generic information about the SAM/BAM file. Each line in the header section starts with the symbol "@". By contrast, the alignment section contains the read name, read sequence, read quality, alignment information, and custom tags. The read name includes the chromosome, start coordinate, alignment quality, and the matching descriptor string. Fig. 10.2 reports an example of an alignment complaint with the SAM format.

```
@HD VN:1.6 SO: coordinate
@SQ SN:ref LN:20
r001 89 ref 7 33 9AG44E3M1N = 37 39 AGTTAGTTAGAG  *
r002 0  ref 8 33 9S2A4M2T2M =  0  0 AAAAGG        *
r003 0  ref 8 33 7A8G       =  0  0 AGTTAGAG      * NM:j:2
```

FIGURE 10.2

Overview of SAM BAM files content.

10.3 Transcriptomics

Transcriptomics [194] indicates the branch of functional genomics that deals with omics technologies to study the transcriptome. The transcriptome refers to the set of

RNA molecules (mRNA, rRNA, tRNA, and other non-coding RNAs) of a cell or organism, with the aim of cataloging them, quantifying them, and characterizing them from a structural point of view to understand their variability factors and functioning mechanisms. The coming of NGS techniques has revolutionized this discipline, characterized by the gradual forsaking of transcripts hybridization techniques based on:

- oligonucleotide probes (gene expression microarray) effective in differential gene expression investigations of known genes;
- RNA-seq [195] more robust approaches (direct sequencing through NGS of cDNA obtained starting from the RNA of interest) able to concurrently enable the exact quantification of the target RNA;
- the identification of the transcription start and stop sites, the ends between exons and introns;
- the identification of possible splicing mRNAs variants (hence, also known as spliceosomal); and
- the quantitative analysis of the expression at the level of the single allele used to highlight possible genetic imprinting phenomena.

Moreover, unlike gene-expression microarrays, RNA-seq techniques make possible identifying novel transcripts, including less abundant transcripts using the coverage score defined in Eq. (10.1).

$$C = totalReads \times \frac{singleReadLength}{wholeGenomeLength} \tag{10.1}$$

10.3.1 RNA-seq data analysis

This section introduces various basic computational procedures involving high-throughput sequence data, which can be achieved or at least managed using an analysis workflow made of various software tools. Although this is a vast and rapidly developing subject, we can introduce the core concepts here to provide solid starting points for further development.

Sequencing methods can generate a massive number of sequence reads for a single experiment. However, no sequencing technology is perfect since each instrument may cause different types and amounts of errors, such as incorrect nucleotide classifications. Hence, sequence quality control is, therefore, an essential step in RNA-seq analysis. Furthermore, sequence quality control is mandatory to figure out and exclude error types that may impact the interpretation of the downstream analysis. The RNA-Seq quality control task can be accomplished using tools like FastQC [196] and Sickle [197] (for quality assessment) and Cutadapt [198] (for trimming and filtering).

Independently of the used technique to generate the DNA segments, the RNA-seq must be mapped to a reference genome to determine where the sequence starts in the chromosomes. Mapping RNA-seq reads back to a genome sequence makes it possible for the reads to be annotated with all the associated genomic information. Usually, discovering where DNA fragments come from requires an alignment of the read sequences to the reference genome sequence to determine where they match. Alignment can be done using tools like Cufflinks (available at https://pypi.org/project/cufflinks/), STAR [199] and StringTie [200] to reconstruct transcripts from spliced read alignments generated by other programs (TopHat [201], TopHat2 [202], HISAT [203], STAR, HTSeq [204]). Other tools, such as Sailfish [205], Kallisto [206], and Salmon [207], perform lightweight alignment of RNAseq reads against existing transcriptome sequences. Finally, the alignment of RNA-seq reads to a genome can be compressed using the Burrows–Wheeler Transform (BWT) [208].

Once reads have been mapped, individual transcripts have to be estimated. RNA-Seq data analysis relies on counting reads generated from various regions of the genome. The total number of reads mapped to a specific genomic region indicates the level of transcriptional activity in the region. The more transcriptionally active a genomic region is, the more copies of RNA transcripts it produces, and the more reads it will generate. Estimating transcript levels can be done using the StringTie software tool.

Thus, RNA-seq analyses are performed to comprehend transcriptomic changes in organisms in response to a specific treatment. They are also intended to understand the mutation cause and/or effect by estimating the resulting gene-expression differences. DESeq2[1] [209] software tool can be used to perform differential gene-expression analysis. Fig. 10.3 shows, for each step of the RNA-Seq analysis pipeline, the recommended software tools.

10.4 Epigenomics

Epigenomics [210] deals with systematically studying the set of epigenetic modifications, i.e., those transmissible, although reversible, modifications capable of altering gene-expression levels without changing the DNA sequence. These mechanisms play an essential role in regulating gene expression. Consequently, they are involved in numerous cellular processes, such as in the ability of a cell or organism to change its state in response to various environmental stimuli. Such mechanisms include chromatin modifications, such as DNA methylation and histone modifications, and RNA interference phenomena. The DNA methylation in the cytosine residues in the CpG islands related to regions with a regulatory function can repress transcription through mechanisms that define a chromatin packing that prevents access to transcription factors. The chemical conversion of unmethylated cytosines into uracil is at the base of DNA methylation. Omics technologies can implement DNA methylation into systems based on microarrays or NGS.

The microarrays comprise two kinds of oligonucleotide probes. The first type of oligonucleotide probe hybridizes specifically to the methylated sequences (e.g.,

[1] http://bioconductor.org/packages/release/bioc/html/DESeq2.html.

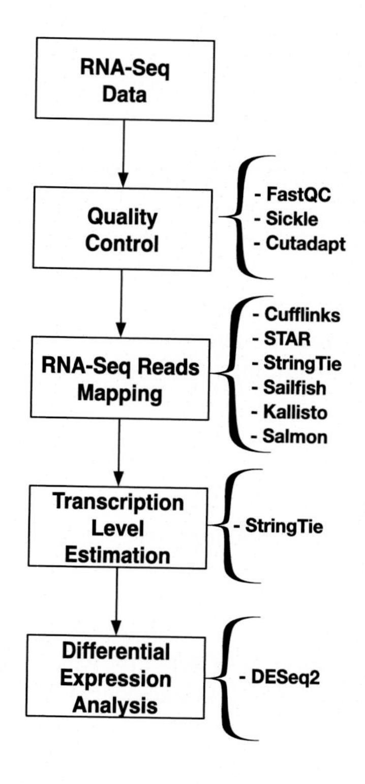

FIGURE 10.3

RNA-Seq data analysis steps and the related software tools.

cytosine residues are unmodified). The second type of oligonucleotide probes can hybridize to the unmethylated sequence (e.g., cytosine residues are converted to uracil by bisulfite treatments). Thus, microarrays are well suited to perform cases-controls studies. Several automatic software tools for statistical and data mining analysis of microarray data sets are available [40,41,211–214]. On the other hand, NGS techniques enable the detection of methylated and non-methylated sites by direct sequencing. Compared to microarrays-based systems, NGS techniques provide higher resolution, better coverage of the genome (e.g., whole-genome shotgun), reliable quantification of methylated sites, and potential use even in species for which the lack of information on the genomic sequence does not allow the design of specific microarrays.

Complementary large-scale methods are available and based, e.g., on immunoprecipitation of methylated DNA (MeDIP) with antibodies directed towards the residues of 5-methylcytosines. The methylated DNA fraction obtained can subsequently be analyzed by microarrays (MeDIP-chips) or NGS (MEDIP-seq). Post-translational modifications of various kinds of histone proteins can influence chromatin structure, with significant repercussions on gene expression and processes, such as DNA repair, replication, and recombination. The analysis of these modifications is commonly carried out at the genome-wide level using chromatin immunoprecipitation (ChIP) coupled with systems based on microarrays (ChIP-on-chip) or by NGS (ChIP-seq). Depending on the antibodies used, these techniques can find more general applications in recognizing protein–DNA binding sites capable of identifying various genome functional elements, such as promoter regions and sequences that control DNA replication. The study of RNA interference (RNAi) phenomena can be approached, at the omic level, by using screening techniques that promote the systematic identification of genes involved in the determinism of specific phenotypes through gene-silencing mechanisms, using various types of synthetic RNA. Genome-wide RNAi screening approaches are suitable for performing in vitro studies.

10.5 Proteomics

Proteomics refers to the structural and functional study of the entire collection of proteins expressed in a specific cell type or organism, called the proteome. The greater level of complexity that characterizes the proteome with respect to the genome and the transcriptome, not only in terms of the number of gene products (at least an order of magnitude higher than the number of genes present in a genome) in terms of variability of expression and localization levels depending on the cell type, considerably complicates the study, despite the development in recent years of better performing technologies than the traditional 2-D electrophoresis combined with mass spectrometry.

10.6 Metabolomics

Metabolomics refers to the identification, quantification, and functional interpretation of endogenous metabolites that are dynamically present in various ways in cells, tissues, and biological fluids.

What follows is a schematic description of the leading investigation methods used in proteomics and metabolomics studies. Among the leading investigation methods, we highlight microarrays, mass spectrometry, and nuclear magnetic resonance.

Microarrays make it possible to investigate on a large scale different forms of interactions involving proteins (e.g., interactome), such as protein–protein interactions, proteins–peptides, enzymes–substrates, receptors–ligands, as well as any opportunities, for example, in the study of signaling pathways, in drug discovery studies, in the understanding of host-pathogen interaction mechanisms, and in the identification of prognostic and diagnostic biomarkers.

Mass spectrometry relies on the separation of ions as a function of their mass–charge ratios. Mass spectrometry technique is available in multiple variants depending on, for example, the combination with preliminary separation techniques, such as capillary electrophoresis (CE), gas chromatography (GC), or liquid chromatography (LC), this latter also in its high-performance variant (HPLC). The methods used to induce the ionization of the molecules to analyze, such as matrix-assisted laser desorption/ionization (MALDI) along with its variants, surface-enhanced laser desorption-ionization (SELDI) or the electrospray ionization (ESI) technique. The type of analyzers used to separate the ions, such as time-of-flight (TOF), ion trap (IT), quadrupole (Q), Fourier-transform ion cyclotron resonance (FT-ICR), and tandem mass spectrometry configuration makes possible to repeat the analysis of fragments already analyzed after subsequent reframing, allowing to reach higher levels of selectivity and sensitivity, particularly helpful in the analysis of complex mixtures. Furthermore, these configurations will enable the definition of the amino acid sequence of a protein and, accordingly, its accurate identification, together with a quantitative evaluation of its presence.

Nuclear magnetic resonance (NMR) is widely used in proteomics and metabolomics. The NMR approach allows, in its most popular version, i.e., proton nuclear magnetic resonance (1HNMR), to analyze biological fluids, cellular extracts, and cell cultures (solution-state NMR), and in the magic-angle spinning (MAS) version, even intact tissues and solid samples (using solid-state NMR). NMR offers some advantages when compared to MS, such as lower cost, simpler sample preparation, and faster, reproducible, and non-destructive analysis. However, NMR has lower sensitivity when compared to MS. Thus, NMR can be used as a complementary approach to MS.

10.7 Interactomics

The complete proteins set in a cell is called the proteome [215], whereas the study of protein structure and function is called proteomics [216]. Moreover, a cell is a highly complex entity in which diverse biomolecules work together in a coordinated fashion to sustain life's biochemical functions. This collective action is accomplished, in large part, by a combination of intermolecular interactions, including protein–protein interactions, protein–DNA interactions, RNA interactions, and many others. Based on that, the proteome is highly dynamic, and it changes from time to time in response to various environmental impulses. Proteomics aims to understand how proteins' structure and function allow them to do what they do, interacting with and contributing to life processes. Biomolecular interactions are conveniently represented as networks (graphs) with nodes (vertices) representing molecules and links (edges) representing interactions among them. Depending on the type of interaction, the corresponding edge might be directed or not. For example, a binding of two proteins is typically represented by an undirected edge, while interaction between a transcription factor and a gene whose expression is regulated by the given transcription factor is usually represented by a directed edge, which can be weighted to reflect its strength. Defining the interactome as a network provides a convenient way to describe an ensemble of objects and their relationships. Thus, a network is entirely represented by the set of its connections, not by how it is drawn. Formally, a graph could be defined as a couple $G = \{V, E\}$, where V is the nodes set and $E = (v_i, v_j)|v_i, v_j \in V$ is an edge between a couple of nodes. A graph G is directed if the edge $(v_i, v_j) \neq (v_j, v_i)$, otherwise, if $(v_i, v_j) = (v_j, v_i)$, G is an undirected graph. Hence, representing biological networks as a graph enables us to analyze various network properties, providing valuable insight into the internal organization of a biological network. A fundamental property to characterize networks is the node degree $\delta(v)$, the number of edges per vertex. Most networks present in nature have large oscillations in the degree value. This feature has profound consequences on the stability of the systems represented by the network and the dynamics of the processes defined on this structure. If (v_i, v_j) is an edge in a graph G between nodes v_i and v_j, the vertex v_i and v_j are adjacent. Thus, the distance between two vertices is called a *path*. The minimum number of edges found between two vertices is called the *shortest path*. Given this rather general definition, it is easy to understand why biological networks are becoming a trendy way to represent a variety of functions in the cell. In particular, interactomics focuses on the modeling, storage, and retrieval of protein–protein interactions (PPI) and algorithms for analyzing protein interaction networks (PIN) or predicting interactions. Since most protein functions are achieved when proteins interact, several experimental techniques to detect them have been developed. These methods can be divided into physical (e.g., concentration, immunoprecipitation), library-based (e.g., protein probing, two-hybrid systems), and genetic methods (e.g., synthetic lethal effects). The two leading technologies used to detect protein–protein interaction are *i)* yeast two-hybrid assay (Y2H) [217] and *ii)* protein complex purification, followed by mass spectrometry (CoIP/MS) [218] identification. The Y2H is a method to reconstruct functional transcription factors (TF) when two proteins or polypeptides of interest interact. The fundamental properties of Y2H are:

1. It detects binary interactions only, and thus might miss interacting proteins that require additional proteins, such as the members of a protein complex;
2. it uncovers the causes of the interaction. For example, two proteins interacting in a cell depend upon spatial or temporal aspects.

The CoIP/MS is a method to determine co-complex identification, including tandem affinity purification (TAP) followed by mass spectrometry TAP/MS [219]. Conversely to the Y2H approach, CoIP/MS can identify one-to-many interactions. In the CoIP/MS approach, a protein bait is tagged with a molecular marker to interact with all other proteins (preys). The interactions among the protein's bait and preys will allow the formation of protein complexes and co-complex proteins, identified using MS. The fundamental properties of CoIP/MS are:

1. The bait protein does not necessarily interact directly with all proteins in the complex;
2. transient interactions, e.g., protein interactions formed and broken easily, that are essential in many aspects of cellular function are usually challenging to identify.

Alongside the experimental methodologies to detect protein–protein interactions, there are computational ones. Computational methods include a variety of techniques, which can roughly be divided into three categories: evolutionary-based approaches, statistical methods, and machine learning techniques. These approaches are designed to predict physical interactions, providing important clues about the functioning of cells and organisms.

10.8 Gene prioritization

The results of high-throughput experiments consist of numerous candidate genes, proteins, or other molecules potentially associated with diseases. A challenge for omics science is knowledge extraction from these candidate lists and the filtering of promising gene or protein candidates. In particular, the hot topic in clinical scenarios consists of highlighting the behavior of the few molecules related to some specific disease. A central question in omics sciences is the analysis of diseases at a molecular level in order to identify genes and molecular markers associated with observed phenotypes. Many experimental platforms, including mass spectrometry and microarray RNA-sequencing, produce vast amounts of data that have to be analyzed to extract valuable knowledge. Specifically, the goal consists of the identification of set of candidate genes potentially associated with the investigated phenotype or involved in analyzed diseases [220,221]. To facilitate the discovered associations in clinical scenarios, different approaches have been developed, also known as *Gene prioritization methods*. They aim to identify the most promising gene candidates on a long list of genes through the use of computational methods [222].

The gene-prioritization task involves the following steps: (1) selecting a set of genes from an high-throughput experiments, i.e., microarrays or NGS, (2) collecting disease information from the literature or from existing knowledge domains, i.e., biological ontologies or repositories, (3) selection of a prioritization method, and (4) identifying the top-ranking genes as the best candidates (see Fig. 10.4).

In recent years, diverse gene-prioritization approaches have been designed and also implemented within publicly available tools. It is possible to classify the existing gene-prioritization methods according to the kind of input data, the used knowledge domains, and the prioritization methods. A central focus in gene prioritization is the integration of heterogeneous data sources related to biological knowledge. In fact, the data sources represent the core of the gene-prioritization approach because the quality of the associations among a disease and candidate genes correlates with the quality of the data used to make these associations. There exist various kinds of data sources, such as molecular interactions, biological pathways, expression, sequence, phenotype, conservation, regulation, disease probabilities and chemical components, gene product protein family domains and functional annotations, and protein–protein interactions. An example of a knowledge source is biological and biomedical ontologies. In fact, the filtering of promising candidate genes requires the use of domain-specific knowledge that is often encoded in ontologies. The most widely used ontologies that

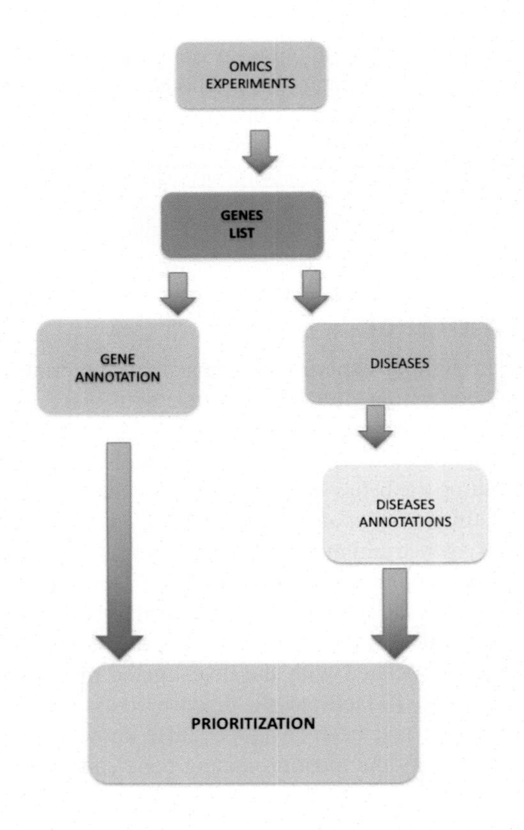

FIGURE 10.4

Gene-prioritization steps.

may improve the prioritization process are Gene Ontology (GO) [223], Human Phenotype Ontology (HPO) [224], and Disease Ontology (DO) [225].

These ontologies ensure the organization of concepts through taxonomies and the association of a concept to gene products through annotations. In general, the ontologies are widely applied only in the post-processing steps of gene prioritization methods in order to mine textual datasources. What's more, various software tools apply gene ontology to help filtering the gene lists on the basis of biological or clinical considerations (e.g., which molecules are more related to a given disease). However, few tools ensure the use of various different ontologies [222]. These tools evaluate the similarity among the annotations of each gene or gene product and of a selected disease by applying semantic-similarity measures. Semantic similarities are formal instruments to quantify the similarity of concepts in a semantic space. Thus, semantic similarities are used to quantify the similarity of a gene to a disease using the annotations of both elements on the same ontology.

10.8.1 Gene-prioritization approaches

In general, the process of prioritization consists of four different steps:

1. First, multiple technologies, such as genomic microarrays, gene-expression microarrays, next-generation sequencing, or micro-RNA microarrays are utilized in an *omics* experiment to identify a list of candidate genes. This list can be filtered using computational techniques, i.e., highlighting differentially expressed genes or potential mutations different in case/control studies.
2. The second step comprises the collection of prior knowledge about the disease. Prior knowledge can be derived from the literature using text-mining tools or as keywords selected by biologists or medical doctors, or it can be derived from ontologies that represent a standardization of knowledge sources.
3. The third step regards the selection of a suitable prioritization method. The correct selection is decisive to achieving the best results.
4. The fourth step consists of the evaluation of obtained results. For example, a post-processing tool can evaluate the biological significance or the enrichment of selected candidate genes.

In recent years, the literature has been supplemented by the implementation of several gene prioritization methods. In general, these tools can differ in terms of the inputs that they require, the methods that they apply, and the outputs they produce. Next, we identify and briefly detail the tools that require annotations for the prioritization calculation.

For example, GeneApart [226], BioGraph [227], and Endeavour [228] take as input a list of gene ontology annotations, and they produce as output a prioritized list of candidates.

MedSim [229] prioritizes a list of genes for a known disease by applying two possible semantic similarity measures, Lin and SimRel, to gene ontology annotations.

ToppGene [230] is a web-based software able to rank candidate genes based on a similarity score for each annotation of each candidate with respect to a selected disease. Finally, the gene prioritization tool, namely GoD (Gene on Diseases) [231], enables the prioritization of a list of input genes with respect to a selected disease. GoD is listed in the Omics Tool Page (https://omictools.com/gene-ranking-based-on-diseases-tool).

In the following, the GoD (Gene on Diseases), a tool that ranks a given set of genes based on ontology annotations, is presented. The GoD algorithm orders genes by the semantic similarity computed with respect to a disease among the annotations of each gene and those describing the selected disease. GoD uses HPO ontology, GO ontology, and DO ontology to calculate rankings. It takes as input a list of genes or gene products annotated with GO Terms, HPO Terms, or DO Terms and a selected disease described in terms of annotation of GO, HPO, or DO. It produces as output the ranking of those genes with respect to the input disease. Fig. 10.5 shows the workflow of the GoD package.

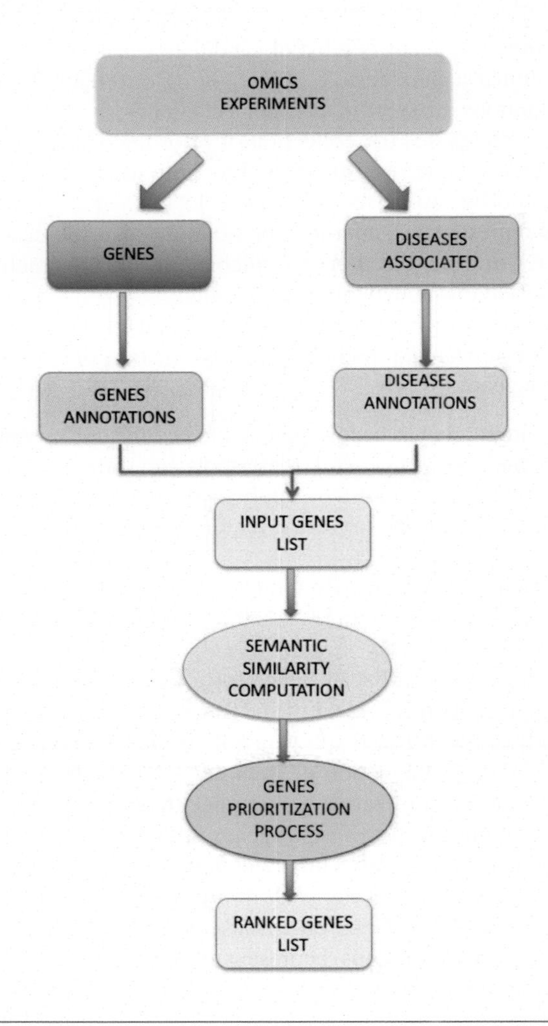

FIGURE 10.5

The workflow of the GoD system.

CHAPTER

Ontologies in bioinformatics

11

CONTENTS

Abstract

Biological data about genes, proteins, and biologically relevant molecules, that are stored in databases, may be associated to biological information (knowledge) such as experiments, properties and functions, response to drugs, etc. Such knowledge is formally structured in ontologies that provide the best formalization to organize and store knowledge. Consequently, it is possible to introduce novel analytical methodologies that are based on the use of ontologies. An example is represented by semantic similarities, i.e., the calculation of the similarity of two or more proteins starting from their annotations. For instance, semantic measures have been used for the prediction of protein complexes. Otherwise, functional enrichment analysis methods ensure the discovery of the functions of the input list of genes by using annotations. This chapter describes the main biomedical ontologies presented in the literature and their bioinformatic applications. Then, the principal semantic similarity measures are presented. Finally, the treatment of functional enrichment analysis concludes the chapter.

Keywords

Biomedical ontologies, Gene ontology, Semantic similarities, Functional enrichment analysis

11.1 Introduction

The production of biological data about genes, proteins, and various relevant molecules stored in databases has led to the production of knowledge, such as exper-

Artificial Intelligence in Bioinformatics. https://doi.org/10.1016/B978-0-12-822952-1.00021-8

iments, properties and functions, and responses to drugs. In general, the biological knowledge is encoded by using special terms to describe the function or localization of genes and proteins. These terms are structured by using formal instruments, such as controlled vocabularies and ontologies, that offer a formal framework for data modeling and representation. Biomedical ontologies are commonly used to structure and organize formal knowledge about biological and biomedical concepts. The terms structured within ontologies are usually associated with biomedical entities in a process referred to as annotation. Important examples of ontologies are Gene Ontology (GO) [232], Human Phenotype Ontology (HPO) [233], and Disease Ontology (DO) [234]. They offer both the organization of concepts through taxonomies and the association of a concept to genes and gene products through annotations. In order to evaluate the similarity among terms belonging to the same ontology, researchers developed a set of formal instruments called Semantic Similarity Measures (SSM) [235,236]. A SSM takes as input two or more terms of the same ontology and produces as output a numeric value representing their similarity. Since proteins are annotated with a set of terms extracted from an ontology, the use of SSMs to evaluate the functional similarity among proteins is becoming a common task. Furthermore, proteomic, genetic, and metabolic analyses has led to the production of biomolecules lists, i.e., lists of genes ranked according to diffenrent co-expression level that are obtained in microarray experiments. In this contest, computational analysis enables the identification of a subset of products with a readable behavior pattern from the initial gene list. In particular, functional enrichment is a computational analysis that enables to read the results from a biological perspective.

11.2 Biomedical ontologies

The acquisition of information during research activities is often hindered by the different terminologies used in scientific fields, which limits the effectiveness of the work of the researchers themselves and of the computers. For example, the search for a new target for an antibiotic requires the investigation of all the gene products involved in the synthesis of bacterial proteins, which are significantly different in sequence or structure from those of humans. However, the databases use different terminologies to describe how molecules are involved in translation so it will be difficult to find terms that are functionally equivalent for both a human and a computer. Advances in technology have made it possible to perform large-scale experiments that yield results for thousands of genes or genetic products in single experiments. The data from these experiments is growing in public repositories, and the focus has shifted from generating this data to analyzing it. In computer science, an ontology refers to a series of representative primitives used to model a knowledge domain. In particular, bioinformatics and computational biology in recent decades have made extensive use of ontologies.

Ontologies are computational structures that describe the entities and relationships of a domain of interest in a structured computational format, which allows their use in multiple applications. An ontology contains set of entities, also called classes, which are organized in a hierarchy from the most general to the most specific. Additional information can be acquired as domain-relevant relationships between entities or even complex logical axioms. The entities that are contained in ontologies are then available for use as hubs around which data can be organized, indexed, aggregated, and interpreted, across multiple services, databases, and different applications.

In computer science, an ontology is therefore a formal, shared, and explicit representation of a conceptualization of a domain of interest. More specifically, it is an axiomatic theory of the first order that can be expressed in a descriptive logic. The term formal ontology has come into use in the field of artificial intelligence and knowledge representation to describe how different schemas are combined into a data structure containing all relevant entities and their relationships in a domain. Computer programs can then use ontology for a variety of purposes, including inductive reasoning, classification, and various problem-solving techniques.

Ontologies are made up of several distinct elements, including classes, metadata, relationships, formats, and axioms. Class is the basic unit within an ontology, representing a type of thing in a domain of interest. Typically, classes are associated with a unique identifier. These identifiers are not related to semantics (they do not contain a reference to the name or definition of the class) in order to promote stability even as scientific knowledge and its ontological representation evolve. These identifiers are also used for annotations or other application contexts. When a class is believed to be no longer needed within the ontology, it can be marked as obsolete, which indicates that the ID should not be used in further annotations, although it is retained for historical reasons. Obsolete classes may contain metadata pointing to one or more alternate classes that should be used instead. In some cases, multiple entries can be merged into one, and the identifiers take the name of secondary identifiers. Classes are generally associated with annotated textual information, i.e., metadata. The metadata associated with the classes can include any secondary identifiers and flags to indicate whether the class has been marked obsolete. It can also include cross-references to that class in alternative databases and web resources. Classes are organized in a hierarchy from general (high in the hierarchy) to specific (low in the hierarchy). Despite the hierarchical organization, most ontologies are not simple trees. Rather, they are structured as direct acyclic graphs. This is because it is possible for classes to have multiple parents in the classification hierarchy, and, in addition, ontologies include additional types of relationships between entities other than the hierarchical classification (which is itself represented by _ relationships). All relationships are direct, and care must be taken to ensure that the general structure of the ontology does not contain cycles.

11.2.1 Gene ontology

Gene Ontology (GO) [232] is one of the main resources of biological information since it provides a specific definition of protein functions. GO is a structured and controlled vocabulary of terms, called GO terms. GO is subdivided in three non-overlapping ontologies: Molecular Function (MF), Biological Process (BP), and Cellular Component (CC). The structure of GO is a Directed Acyclic Graph (DAG), where the terms are the nodes and the relationships among terms are the edges. The structure allows for more flexibility than a hierarchy since each term can have multiple relationships to broader parent terms and more specific child terms [237]. Genes or proteins are connected with GO terms through annotations by using a procedure also known as an annotation process. Each annotation in the GO has a source and a database entry attributed to it. The source can be a literature reference, a database reference, or computational evidence. Each biological molecule is associated with the most specific set of terms that describe its functionality. Then, if a biological molecule is associated with a term, it will connect to all the parents of that term [237]. *Fourteen* different annotation processes exist that are identified by an evidence code, the principal attribute of an annotation. The available evidence codes describe the basis for the annotation. A main distinction among evidence codes is represented by ones known as Inferred from Electronic Annotations (IEA), i.e., annotations that are determined without user supervision, and non-IEA ones or manual annotations, i.e., annotations that are supervised by experts. The current GO version (January 2022) contains 43,789 terms and 7,877,711 annotations.

11.2.2 Human phenotype ontology

Human Phenotype Ontology (HPO) [233] is a standardized, controlled vocabulary that contains phenotypic information about genes or product genes. The HPO currently contains over 10,088 terms describing phenotypic features. The ontology is organized as five independent ontologies that include different categories: *Mode of inheritance* that indicates disease models according to Mendelian or non-Mendelian inheritance modes; *Mortality* that records the age of death associated with a disease; *Clinical modifier* that contains terms to characterize and specify the phenotypic abnormalities defined in the phenotypic abnormality sub-ontology; *Frequency* that indicates the frequency of clinical feature display from patients; and *Phenotypic Abnormalities* that contains terms describing organ abnormalities. The HPO is structured into a directed acyclic graph (DAG) in which the terms represent subclasses of their parent term and each term in the HPO describes a distinct phenotypic abnormality. HPO currently contains over 13,000 terms and over 156,000 annotations to hereditary diseases that are available at the website (http://www.human-phenotype-ontology.org). Diseases are annotated with terms of the HPO, meaning that HPO terms are used to describe all the signs, symptoms, and other phenotypic manifestations that characterize the disease in question. Since HPO contains information related to phenotypic abnormalities, the computation of semantic similarities among concepts annotated with HPO terms may enable database searches for clinical diagnostics or computational analysis of gene expression patterns associated with human diseases.

11.2.3 Disease ontology

The Disease Ontology (DO) database [234] (http://disease-ontology.org) is a knowledge base related to human diseases. The current version stores information about 15,000 DO terms. The aim of DO is to connect biological data (e.g., genes) from a disease-centered point of view. The DO semantically integrates a disease and existing medical vocabularies. DO terms and their identifiers (DOIDs) have been utilized to annotate disease concepts in several major biomedical resources. The DO is organized into eight main nodes to represent cellular proliferation, mental health, anatomical entity, infectious agent, metabolism, and genetic diseases along with medical disorders and syndromes anchored by traceable, stable DOIDs. Genes may be annotated with terms coming from DO, which may be freely downloaded from the website. DO is structured into a directed acyclic graph (DAG), and the terms are linked by relationships in a hierarchy organized by interrelated subtypes. DO has become a disease knowledge resource for the further exploration of biomedical data, including measuring disease similarity based on functional associations between genes, and it is a disease data source for the building of biomedical databases.

11.3 Semantic similarity measures

One of the main purposes of the annotations is to allow the quantitative comparisons of gene functions based on the measure of functional similarity between two genes, defined by the GO terms, associated with these genes. In the scientific field, comparison and classification are widely used methods; scientific knowledge, biological laws and models derive from the analysis of similarity and differences between genes, cells, organisms, populations, and species. However, since biology can hardly represent knowledge through mathematical formulas, it has been used either by natural language for publications or by classification schemes (a new knowledge is compared with known entities to deduce the level of similarity or difference). The comparison of entities is not always trivial; while, for example, the sequences or structures of two genes products can be directly compared through the use of alignment algorithms, this cannot be done for the functional aspect because it is not objectively measurable. This need has led to the development of ontologies with the aim of annotating gene products (gene ontology), sequences (sequence ontology) and experimental analyzes (microarray and gene-expression data ontology). The ontologies, through the annotations, make possible the comparison of entities for aspects that are difficult to compare by other means, such as, for example, comparing semantic similarity measures in a set of interactions of gene products that have common terms.

An SSM is a formal instrument to quantify the similarity of two or more terms in the same ontology. Measures comparing only two terms are often referred to as pairwise semantic similarity measures, while measures that compare two sets of term yielding a global similarity among sets are referred to as groupwise measures. The pairwise approach measures the functional similarity between two gene products by combining the semantic similarities between their terms. Each gene product is represented by a group of direct annotations, and the semantic similarity is calculated between the terms of one set and the terms of another, using one of the previously described methods for comparing terms. Some approaches consider all combinations in pairs of the terms of the two sets—the all-pairs technique—while others consider only the best matching pairs for each term—the best-pairs technique. A global functional similarity score between gene products is obtained by combining these semantic similarities in pairs, with the combination of the best commonly used methods such as the average, the maximum, and the sum.

Groupwise approaches to calculate the similarity of the gene product are not based on the combination of similarity between single terms but rather one of the following three approaches: set, graph, or vector. In the set method, only direct annotations are considered, and functional similarity is calculated using a set of similarity techniques. In the graph method, the gene products are represented as GO subgraphs corresponding to all their annotations; functional similarity can be calculated using graph matching techniques or considering subgraphs as term sets and applying sets of similarity techniques. In the vector method, a gene product is represented as a vector space in which each term corresponds to a dimension, and functional similarity is calculated using vector similarity measures. Vectors can be binary, where each dimension denotes the presence or absence of the term in the annotation set of a given dimension; each dimension represents a given property of the term (e.g., its Information Content (IC)). Since proteins and genes are associated with a set of terms coming from gene ontology, $SSMs$ are often extended to proteins and genes. The similarity of proteins is then translated into the determination of similarity of a set of associated terms [235,236]. Many similarity measures have been proposed (see, for instance, [238] for a complete review) that may be categorized according to various strategies used for evaluating similarity.

For instance, many measures are based on **Information Content (IC)** of ontology terms. The IC of a term T of an ontology O is defined as $-\log(p(c))$, where $p(c)$ is the number of concepts that are annotated with T and its descendants, divided by the number of all concepts that are annotated with a term of the same ontology. Measures based on a **common ancestor** first select a common ancestor of two terms according to its properties and then evaluate the semantic similarity on the basis of the distance between the terms and their common ancestor and the properties of the common ancestor. IC can be used to select the proper ancestor, yielding to the development of mixed methods.

The Resnik's similarity measure sim_{Res} of two terms T_1 and T_2 of GO is based on the determination of the IC of their most informative common ancestor ($MICA$), where $MICA$ is the common ancestor with the highest IC [239]:

$$sim_{Res} = IC(MICA(T_1, T_2)). \tag{11.1}$$

A drawback of the Resnik's measure is that it considers mainly the common ancestor, and it does not take into account the distance between the compared terms and the shared ancestor. Lin's measure [240], sim_{Lin}, addresses this problem by considering both terms and yielding to the formula:

$$sim_{Lin} = \frac{IC(MICA(T_1, T_2))}{IC(T_1) + IC(T_2)}. \tag{11.2}$$

Jiang and Conrath's measure sim_{JC} takes into account the distance between terms by calculating the following formula:

$$sim_{JC} = 1 - IC(T_1) + IC(T_2) - 2 * IC(MICA(T_1, T_2)). \tag{11.3}$$

Completely different from previous approaches are the techniques based on **path length** (edge distance). In this case, similarity measures are correlated to the length of the path connecting the two terms. Finally, many integrative approaches of two categories have recently been proposed to achieve higher accuracy in measuring functional similarity of proteins. For example, Wang et al. [236] proposed a combination of the normalized common-term-based method and the path-length-based method. Their semantic similarity measure scores a protein pair by the common GO terms having the annotations of the proteins but gives different weights to the common GO terms according to their depth.

Pesquita et al. [241] proposed simGIC that integrates the normalized common-term-based method with information contents. Instead of counting the common terms, simGIC sums the information contents of the common terms:

$$sim_{simGIC}(T_1, T_2) = \frac{\sum_{T_i \in C(T_1) \cap C(T_2)} \log P(T_i)}{\sum_{T_j \in C(T_1) \cup C(T_2)} \log P(T_j)},$$

where $C(T_1)$ is a set of all ancestor terms of T_1. Finally, two recent IC based measures were proposed by Cho et al. [242].

simICNP (Information Content of SCA Normalized by Path-length of two terms) uses the information content of common ancestors normalized by the shortest path length between T_1 and T_2 as the distance:

$$sim_{simICNP}(t_1, t_2) = \frac{-\log P(T_0)}{len(T_1, T_2) + 1}, \tag{11.4}$$

where T_0 is SCA of T_1 and T_2. This method gives a penalty to Resnik's semantic similarity if T_1 and T_2 are located farther from their SCA.

simICND (Information Content of SCA Normalized by Difference of two terms' information contents) employs the information content of SCA normalized by the difference of information contents from the two terms to SCA, as Jiang's method does:

$$sim_{simICND}(T_1, T_2) = \frac{-\log P(T_0)}{2 \cdot \log P(T_0) - \log P(T_1) - \log P(T_2) + 1}. \tag{11.5}$$

This method gives a penalty to Resnik's semantic similarity if the information contents of T_1 and T_2 are higher than the information content of their SCA. Table 11.1 summarizes the main SSMs.

Table 11.1 Semantic similarity measures.

Similarity Measure	Formula	Ref
sim_{Res}	$IC(MICA(T_1, T_2))$	Resnik
sim_{Lin}	$\frac{IC(MICA(T_1,T_2))}{IC(T_1)+IC(T_2)}$	Lin
sim_{JC}	$1 - IC(T_1) + IC(T_2) - 2 * IC(MICA(T_1, T_2))$	Jiang and Conrath
sim_{simGIC}	$\frac{\sum_{T_i \in C(T_1) \cap C(T_2)} \log P(T_i)}{\sum_{T_j \in C(T_1) \cup C(T_2)} \log P(T_j)}$	Pesquita et al.
$sim_{simICNP}$	$\frac{-\log P(T_0)}{len(T_1,T_2)+1}$	Cho et al.
$sim_{simICND}$	$\frac{-\log P(T_0)}{2 \cdot \log P(T_0) - \log P(T_1) - \log P(T_2) + 1}$	Cho et al.

11.3.1 Knowledge extraction from ontologies

An important bioinformatics research area focuses on the analysis of annotated data in order to extract relevant knowledge. The literature reports various methods for the analysis of annotated data; among these the association rules (ARs) have been applied to analyze a particular aspect of the annotations that concerns the consistency. The need to improve the consistency of the annotations arises from a periodic updating of the ontologies reflecting the improvement of the knowledge about biological and medical phenomena due to the constant efforts of the community of researchers. In contrast, the updating of ontology and annotations carried out by manual curators is slower compared to the rate of introductions of novel facts. The development of ontologies is helped by computational approaches that guarantee remarkable speed. Nevertheless, the use of such methods causes the introduction of possible errors and inconsistencies. Consequently, the need arises to introduce methodologies and tools to support curators to improve the corpus of annotations and the structure of the ontology. Multiple works [243], [244], [245], [246] have proved that ARs can be used to improve annotation consistency by discovering relevant associations. Accordingly, various different tools for learning association rules are being developed to check annotation consistency in biomedical ontologies. An example is GO-WAR (Gene Ontology-based Weighted Association Rules), a novel data-mining approach able to extract weighted association rules starting from an annotated dataset of genes or gene products. GO-WAR [245], [246] calculates the information content for each GO term and extracts weighted association rules by using a modified FP-Tree-like algorithm able to deal with the dimension of classical biological datasets. In particular, GO-WAR is able to handle data coming from the three GO sub-ontologies MF, CP, and BP. GO-WAR is able to manage multiple ontologies and this makes it suitable to analyze data coming from a single or multiple ontologies. In this way, it is possible to highlight unknown relationships among terms belonging to different ontologies or unknown relationships among terms belonging to the same ontology. GO-WAR is publicly available at https://sites.google.com/site/weightedrules/.

Another tool is HPO-Miner [247], a novel framework to learn association rules from HPO. The framework is based on a multithreaded tool able to learn rules in an efficient way. HPO-Miner is based on two main steps: (i) an initial step in which a multithreaded approach is used to calculate the support for the transactions; and (ii) a second sequential step in which an FP-Tree-like structure is used to select relevant rules.

Finally, WARDO [248] is a tool for learning weighted association rules from the DO ontology. WARDO is based on a customized version of FP-Growth algorithm able to deal with DO dataset annotated with IC values. WARDO is able to check annotation consistency and to identify hidden relationships between human disease terms from DO. WARDO is publicly available at https://gitlab.com/giuseppeagapito/wardo.

11.4 Functional enrichment analysis

The advent of high-throughput technologies, such as microarrays and next-generation sequencing, has allowed the screening of huge number of genes and proteins. In fact, bioinformatics analysis applied on initial gene lists has ensured the identification of a subset of products with readable behavior patterns. A task of bioinformatics analysis, also known as functional enrichment, enables the interpretation of results from a biological perspective. In particular, functional enrichment is the process of identifying the functions of the input genes list through the annotations, i.e., the known functions of the genes stored into biological ontologies. In the literature, there are various algorithms and tools invoked for functional enrichment analysis [249]. Furthermore, researchers have developed methodologies for the classification of existing methods and a website with the goal to suggest the use of the most suitable algorithms. An interesting classification was proposed by [249] that categorized the algorithms in three main classes according to the schema applied to analyze input data. In general, the algorithms for functional enrichment analysis take as input a list of genes or proteins and a set of annotations describing the functions, based on a biological ontology (i.e., gene ontology [250]) and then test the relevance of the annotations in the input list by applying statistical or semantic-based methods [238].

The first class consists of Singular Enrichment Analysis (SEA) [251]. This class includes the algorithms that take as input a set of genes or proteins produced by the user, and, then, they iteratively test the enrichment of the annotation terms extracted from an ontology for each gene or protein. After this step, each annotation term is associated with an enrichment probability, also known the enrichment P-value. The P-value threshold measures the probability that genes of the input list have a given annotation term with respect to a random chance. The enrichment P-value can be computed by using well-known statistical methods such as chi-square, Fisher's exact test, binomial probability, and hypergeometric distribution. Finally, the algorithms select the enriched annotation terms that satisfy the enrichment P-value. Although, SEA algorithms demonstrate simple approach and efficiency, their weakness is related to the results. In fact, the set of annotation terms that passes the enrichment threshold may be very large.

The second class is represented by Gene Set Enrichment Analysis (GSEA). The GSEA tools differ from SEA in the calculation of the enrichment P-values [231]. GSEA algorithms take into account all the genes of an experiment (while SEA selects only user selected genes) by reducing the arbitrary factors in the typical gene selection step. Then, they use all the genes to calculate a maximum enrichment score (MES), and genes are ranked. Then, enrichment P-values are calculated by matching the MES to randomly shuffled MES distributions.

The third class consists of Modular Enrichment Analysis (MEA) [251]. MEA computes the enrichment using network-based algorithms [252]; in this way they ensure determination of term-to-term relationships, i.e., the biological meaning of the set of terms.

CHAPTER

Integrative bioinformatics 12

CONTENTS

Abstract

In bioinformatics data integration is the process of merging heterogeneous data produced by multiple laboratories or stored in different databases. More recently, the use of deep learning to mine integrated data (e.g., in an omics field) has gained major attention. We here introduce the main concepts related to data integration, and we present some existing approaches.

Keywords

Data integration, Big data, Data modeling, Data mining equations

12.1 Introduction

The term data integration usually refers to the set of theoretical methods and the implementation tools that aim to combine heterogeneous data coming from multiple sources to provide users a unified view of them. In a commercial scenario, data integration may refer to the merging of the databases of two or more companies. In bioinformatics data integration is generally used as the process of merging many research results produced by multiple laboratories or stored in different databases.

Data integration, in the early steps of its introduction, was the focus of many theoretical work that lead to creating a solid theoretical framework. In particular, the theory of data integration is a subset of database theory based on the formalization through first-order logic. The application of theoretical results sheds light on the data integration in all the real cases.

Artificial Intelligence in Bioinformatics. https://doi.org/10.1016/B978-0-12-822952-1.00022-X

12.2 Data integration in bioinformatics

The use of novel technological platforms for the analysis of molecular insights has produced a large amount of data about multiple aspects of the *omic* world [253]. The recent trend is this area is to integrate multiple data sources, and therefore there is the need for novel approaches and methods to manage, store, and analyze this data [254–256]. This requires the synergistic cooperation among researchers in computer science, bioinformatics, and mathematical modeling for the interpretation of large datasets belonging to different data sources [257,258].

The flow of information in this field starts from the technological platforms that produce various data about molecular biology as depicted in Fig. 12.1.

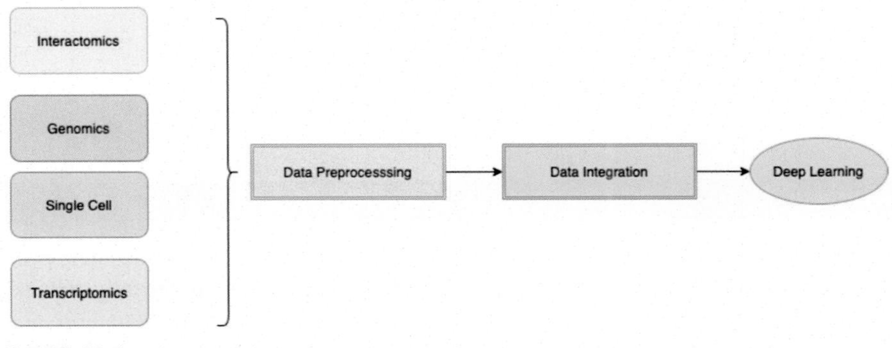

FIGURE 12.1

Data Integration. Flowchart depicts the flow of data in this field. Users may use different samples and different technological platforms to produce their own data. In parallel, previous wet lab experiments or computational experiments (e.g., prediction algorithms) have produced and made the available multiple knowledge banks. The network-based integrated analysis takes as input both user experimental data and knowledge banks and produces biologically meaningful knowledge using appropriate theoretical models.

12.2.1 Deep learning and data integration in multiomics

Multiomics (also known as integrative omics) is a recent trend of biological analysis that relies on the integrated analysis of multiple omics data, such as those related to the genome, proteome, transcriptome, epigenome, metabolome, and microbiome. Such kinds of analysis are often associated with single-cell datasets. Multiomics, from a computer science point of view, is based on the integration of diverse omics data [259,260] as depicted in Fig. 12.2.

The integration of multiomics data has some particular challenges [261] as we briefly review in the following.

First, it should be noted that **multiomic data are often heterogeneous** considering both syntax and semantics. Transcriptomics and proteomics use different

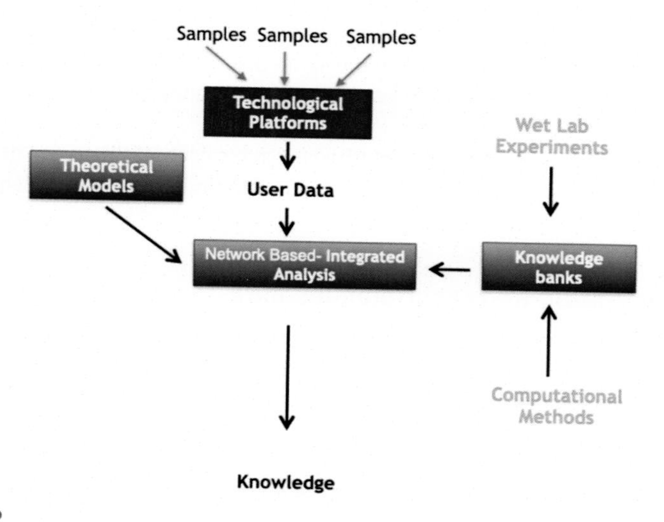

FIGURE 12.2

Data integration: figure depicts the flow of data in this field.

normalization and scaling techniques before the omics analysis if performed. Therefore, it is crucial to perform data pre-processing for each omic dataset before the integration step.

Second, **classes are always unbalanced** due to obvious reasons: Diseases are sometimes rare, and the number of diseased individuals is usually low. Thus, various efforts produce data tests of *sample* individuals to provide class balancing.

Similar to single omics data, multiomics datasets suffer from the classical curse of the dimensionality problem: They contain many fewer samples (n) than features (p). Therefore, it is crucial to use appropriate feature selection and extraction algorithms.

Finally, it is important to ensure transparency and explainability of the whole process since a transparent and explainable data analysis process seems essential to building the trust critical to clinical decision making.

There exist many approaches for the integration of multiomics data that we categorize in the following paragraphs, as presented in [261].

Concatenation-based integration methods receive as input many data sets encoded into various matrices. Each row represents a feature (e.g., a gene) and each column represents a sample. These methods develop a single model using a joint data matrix that is formed by combining multiple omics datasets. The integration is obtained by concatenating multiple matrices to form a single large matrix of multiomics data. Finally, the joint matrix is used for supervised or unsupervised analysis.

Model-based integration methods are based on the creation of an intermediate model for each set of omics data. Then, these intermediate models are built to create a single final model. This integration method has the advantage of being effective in merging models based on different omic types, where each model is developed from a different patient group but containing the same disease information.

Deep learning methods have also been adopted for model-based supervised learning [261].

Transformation-based integration methods reshape each of the omics datasets into graphs or kernel matrices and then combines all of them into one before constructing a model.

12.2.2 Integration of transcriptomics data

Here we focus on the study of the complex mechanisms of the regulation of gene expression. Recent results have confirmed that the transcription of mRNA into proteins is a multi-step process in which various molecules play a synergistic role [262]. In particular, miRNAs and transcription factors (TFs) play a direct role in the regulation of gene expression that results in variable levels of gene transcripts and proteins [263,264]. The integration of this data is made by using a formalism that comes from graph theory.

The approaches discussed here use an internal knowledge base of associations extracted from the literature and databases modeled as a comprehensive graph. The nodes of these graphs are miRNAs, mRNAs, and TFs, while the edges represent activation and inhibition, respectively, among molecules. A directed activation/inhibition edge connects a molecule that increases/decreases the level of another one. The main differences between the approaches are represented by the association databases that are used. This internal knowledge base is used to guide the analysis of experimental data.

Experimental data are miRNA and mRNA and NGS expressions. The graph is built in the following way. There exists a node for each mRNA m_i and each miRNA m_{ij}. Then, the analysis of the expression using some relatedness measures, such as Pearson correlation $\rho(m_i, m_{ij})$ for each mRNA-miRNA pair, produces the weighted edges among them.

Here we report some available webservers and software for the integration of these data.

dChip-GemiNI (Gene and miRNA Network-based Integration) [265][1] is a web server freely available for academic users able to integrate and analyze paired miRNA–mRNA expression data.

To analyze data, the user has to upload into the software two vectors of expression levels (one for mRNA and one for miRNA). Data may be paired (i.e., for each sample there exist both mRNA and miRNA) or non-paired, i.e., data belong to the same class but not to the same samples. Then, the user has to upload them to the web server, and he/she receives as output a list of significant Feed Forward Loops that are altered with respect to those used as the null model.

---

[1] http://www.canevolve.org/dChip-GemiNi.

MAGIA^2 [266][2] is the evolution of the MAGIA web tool for the integrated analysis of both genes and microRNA. MAGIA^2 is available as a free-access web server.

The user has to upload data, usually a matrix for gene/transcripts and one for miRNA expression data, to the web server. The user may integrate both time series experiments or a two-class experiment. The integration among this data may be done using many relatedness measures, e.g., Pearson linear correlation, Spearman rank-based correlation, and an association measure based on information theory for time-series.

The user may also choose which databases are used to extract associations from those explained so far. In the case of the choice of multiple databases, results of search may contain the union or their intersection. Finally, the experimentally derived associations are compared to those contained in the databases, and two kinds of networks are derived.

mirConnX [267][3] is based on a broader perspective with respect to the previous approaches since it uses a genome-wide approach. Unfortunately, it enables the analysis of data of only two organisms: humans and mice. The workflow of analysis is based on the comparison of two networks of associations among genes, TFs, and mRNAs.

The latter first uses a network to derive a null model given by the analysis of databases and literature. The second network, built from experimental data uploaded by the user is obtained by analyzing all the possible pairwise interactions between TFs, miRNAs, and genes across the samples/replicates. The user may choose different measures of associations, both parametric and non-parametric (e.g., Pearson, Spearman, and Kendall).

IntegraMiR [268] is a novel approach for data integration. It receives as input data mRNA and miRNA expressions obtained from samples that are subdivided in two classes (e.g., controls vs. cases). It starts by searching for differences in the two conditions as expressed by genes and mRNAs between two conditions by using the Bioconductor package LIMMA [269].

12.3 Databases, tools, and languages

One of the effects of the big data trend in bioinformatics (i.e., the huge production of omics and clinical data) is the building of novel databases containing omics and clinically curated data [270]. An important initiative in cancer research is The Cancer Genome Atlas (TCGA),[4] a multi-institutional innovative research program supported by the National Cancer Institute (NCI) and the National Human Genome Research Institute (NHGRI).[5] It aims to accelerate understanding of the molecular basis of

[2] http://gencomp.bio.unipd.it/magia2/start/.
[3] http://www.benoslab.pitt.edu/mirconnx.
[4] https://www.cancer.gov/about-nci/organization/ccg/research/structural-genomics/tcga.
[5] https://cancergenome.nih.gov/.

cancer and the discovery of prognostic and predictive biomarkers to improve cancer prevention, diagnosis, and treatment. Recently, TCGA has been superseded by the Genomic Data Commons (GDC),[6] the primary point of access for all NCI datasets, including data from TCGA and from and its pediatric equivalent, Therapeutically Applicable Research to Generate Effective Treatments (TARGET).[7]

Despite the availability of many tools for querying and analyzing TCGA, they are often hard to use, and the integration across data sets and data types remains limited [271]. Moreover, researchers have to build complex pipelines by combining ad hoc scripts and several complementary tools [272]. In recent years, intuitive and user-friendly tools and interfaces have been developed for combining the genomic and clinical data sourced from the TCGA databases discussed so far. Such tools can be divided in two main categories: "Extraction tools" (ET), i.e., tools mainly used to download cancer genomics data, and "Integrative data analysis" (IDA) tools that support integrative data analysis of cancer genomics data, i.e., the analysis of heterogeneous types of data coming from several platforms.

The Cancer Genome Atlas (TCGA)

The Cancer Genome Atlas (TCGA) project is a multi-institutional research program supported by the National Cancer Institute (NCI) and the National Human Genome Research Institute (NHGRI), of the National Institutes of Health (NIH). TCGA contains genomic, transcriptomic, epigenomic, and proteomic data about several cancer types. Moreover, pathology data, slide images for histopathology, and details on patients information are reported for each tumor type [273].

The Cancer Genome Characterization Initiative (CGCI)

The NCI Cancer Genome Characterization Initiative (CGCI)[8] is a comprehensive database of genomic alterations in tumors. It includes next-generation sequencing (NGS) data for genomic characterization (e.g., exome sequencing and transcriptome analysis). To guarantee the privacy of participants, CGCI provides data using two tiers, "open access data" (i.e., high-level genomic information so that patients cannot be identified), or "controlled access data" (i.e., individually identifiable data).

Therapeutically Applicable Research to Generate Effective Treatments (TARGET)

The NCI Therapeutically Applicable Research to Generate Effective Treatments (TARGET) aims to characterize alterations in both gene expression and in genomic structure that can be involved in childhood cancers. Similar to TCGA and CGCI, even TARGET datasets are available to the scientific community as "open access" or "controlled access", depending on the identifiability of the participants by the data.

---

[6] https://portal.gdc.cancer.gov/.

[7] https://ocg.cancer.gov/programs/target/overview.

[8] https://ocg.cancer.gov/programs/cgci.

Genomic Data Commons (GDC) Data Portal

The NCI Genomic Data Commons (GDC) program makes available a huge amount of cancer genomic and clinical data and represents one of the largest and most comprehensive cancer genomics datasets. It is the primary point of access to several large-scale NCI programs, such as TCGA and TARGET, and supports the integration of datasets coming from other databases such as the Cancer Genome Characterization Initiative (CGCI), the Cancer Cell Line Encyclopedia, and the Human Cancer Models Initiative (HCMI).

cBioPortal

The GDC cBioPortal[9] [274] is an open-access web application that supports exploration, integration, and analysis of cancer genomic data, mainly from TCGA and ICGC. The Onco Query Language (OQL)[10] can be used to specify several types of genetic alterations for study within the cBio portal. The OQL can be used in single-cancer and cross-cancer queries. All available data can be downloaded in a tabular format for further analysis. The cBioPortal will soon be replaced by the GDC Data Analysis, Visualization, and Exploration (DAVE)[11] tools.

ICGC Data Portal

The International Cancer Genome Consortium (ICGC)[12] aims to coordinate 80 projects across five continents with the aim to generate a vast catalog of genomic abnormalities in cancer for the study of the biological nature of malignant processes and to develop new targets and biomarkers. The ICGC Data Portal enables researchers to dynamically query complex genomics data by interrogating external databases (e.g., Ensembl, Reactome, and COSMIC). One popular feature of the portal makes it possible to specify three main options: donors, genes, and mutations. The donors option enables filtering the results by the donor's attributes (e.g., gender, vital status, age at diagnosis); the genes option enables filtering the results by a single or a set of genes as input; the mutations option enables filtering the results by a mutation's attributes (e.g., type of mutation).

NCI Cloud Resources

The NCI Cloud Resources[13] is a program for supporting the large-scale analysis of cancer genomic data on the cloud. It allows users to access large genomic datasets without downloading them and to execute bioinformatics pipelines within the cloud. Tools can be combined in workflows through the Common Workflow Language

---

[9] https://www.cbioportal.org/.

[10] https://www.cbioportal.org/oql.

[11] https://gdc.cancer.gov/support/gdc-webinars/gdc-data-analysis-visualization-and-exploration-dave-tools.

[12] https://dcc.icgc.org/.

[13] https://datascience.cancer.gov/data-commons/cloud-resources.

(CWL)[14] or the Workflow Description Language (WDL)[15] that are then executed in the cloud.

TCGA2BED

TCGA2BED [275] is a Java application that allows extracting, extending, and integrating genomic and clinical data from TCGA. It integrates TCGA datasets with annotations (e.g., gene coordinates) retrieved from other genomic databases, such as Entrez Gene and miRBase.

GenoMetric Query Language (GMQL)

The GenoMetric Query Language [276] is a declarative query language for the integration of genomic and clinical data from heterogeneous sources. It is based on the Genomic Data Model (GDM)[16] which combines genomic region data and their associated metadata. GMQL makes it possible to perform massive operations on genomic regions working with distances inside the genome (i.e., region-relative positions and distances). GMQL software is available at [276] for noncommercial use, and it can be executed locally or within a cloud-based environment. It provides a web interface, some REST APIs that access the GMQL repository, the PyGMQL python package and the RGMQL R package.

---

[14] https://www.commonwl.org/.

[15] https://openwdl.org/.

[16] https://docs.gdc.cancer.gov/Data/Data_Model/GDC_Data_Model/.

CHAPTER

Biological networks analysis

13

CONTENTS

Abstract

Various biological systems are modeled according to graph-theory formalism that enables one to represent the entities of system as nodes and their relationships as edges. In general, the biological networks may be categorized as metabolic networks, gene regulatory networks, and protein–protein interactions (PPI). A common analysis for the comparison of graphs is based either on comparing their global properties, such as the clustering coefficient and node-degree distribution, or on the analysis of their internal structure, formally known as network alignment (NA). Furthermore, motif discovery has gained considerable interest because motif analysis has revealed the existence of subsequences that plays important biological roles.

However, efforts are recently focused on novel approaches for encoding structural information about the graph, also known as representation learning. These approaches are able to learn a mapping for nodes (or subgraphs) as points of a low-dimensional vector space.

The chapter introduces the biological networks, such as protein interaction networks, and the main methods developed for motif discovery and network comparison. Also, some network embedding algorithms are presented.

Artificial Intelligence in Bioinformatics. https://doi.org/10.1016/B978-0-12-822952-1.00023-1

Keywords

Network analysis, Network alignment, Statistical analysis, Network embedding, Complex disease

13.1 Introduction

High-throughput technologies are producing an increasing volume of data that needs large data storage, effective data models, and efficient, possibly parallel, analysis algorithms. While proteomics and genomics data, represented as data streams or data tables, are mainly used to screen large populations in case-control studies (e.g., for the early detection of diseases see, for instance, [38,277–280]), interactomics data are represented as graphs and add a new dimension of analysis, enabling, for instance, the graph-based comparison of an organism's properties [255,281].

The modeling of biological systems as graphs is becoming a substantial field of research in bioinformatics and systems biology [282,283]. According to the formalism coming from graph-theory, nodes of the graph represent biological entities, while edges represent the associations among them. Since the study of associations in a system-level scale has shown great potential, the use of networks has become a de facto standard and the application fields span from molecular biology to connectome analysis [284].

For instance, biological networks, also referred to as Protein–Protein Interaction Networks (PINs), model biochemical interactions among proteins [285]. Nodes represent the proteins from a given organism, and the edges represent the protein–protein interactions. Similarly, the graph-based modeling of the whole system of the brain elements and their relationships, the so-called brain connectome [286], is based on the representation of regions of interest (ROI) as nodes, and the representation of functional or anatomical connections as edges.

In both cases, the use of graph theory enables the straightforward analysis of biological properties through the investigation of graph properties.

For instance, the comparison of Protein–Protein Interactions (PPIs) networks has found evidence of the conservation of patterns of interactions in evolution [287,288]. What's more, graph analysis applied to brain connectome has led to the identification of changes in the structure of the networks related to aging and diseases [242,289]. Comparison of graphs is based either on comparing their global properties, such as clustering coefficient or node-degree distribution, or on the analysis of their internal structure, formally known as **network alignment** (NA). NA is based on graph or subgraph isomorphism, and, for PPINs, it is the counterpart of sequencing for the structure of proteins [290,291].

13.2 Networks in biology

Molecules of various types, e.g., genes, proteins, ribonucleic acids, DNA, and metabolites, have fundamental roles in the mechanisms of the cellular processes. The study of their structure and interactions is crucial for various reasons, supporting the development of new drugs and the discovery of disease pathways. For these reasons, the modeling of the complete set of interactions and associations among biological molecules is a burgeoning area of research in bioinformatics and systems biology. In fact, biological molecules have a dense set of associations among them that need to be modeled and represented to improve knowledge in molecular biology.

The best formalism to represent such a complex world comes from graph theory [292], which provides an integrated way to look into the dynamic behavior of the cellular system through the interactions of its components. According to this formalism, many biological processes can be modeled with the entities as nodes and the interactions or relationships among them as edges. The representation as a graph is convenient for a variety of reasons. Networks provide a simple and intuitive representation of heterogeneous and complex biological processes. Moreover, they facilitate modeling and understanding complicated molecular mechanisms through the use of graph theory, machine learning, and deep learning techniques. The biological networks may be categorized at different levels of detail [293] as metabolic networks [294], gene regulatory networks [295], and protein–protein interaction (PPI) [296].

Metabolic networks [294] are used to represent all chemical reactions that involve the metabolism of small molecules, i.e., metabolites, along with the catalytic proteins, i.e., enzymes. Usually, the metabolic networks are decomposed into metabolic pathways that represent a series of chemical reactions related to the execution of a specific metabolic function. In a metabolic network, the nodes represent the metabolites, and the edges represent the reactions. Also each edge is labeled with the enzyme acting as the catalyst.

A Gene Regulatory Network (GRN) [295] is a collection of genes in a cell that interact with each other and with other substances in the cell, such as proteins or metabolites, thereby governing the rates at which genes in the network are transcribed into mRNA. Mathematically, a GRN can be represented as a directed graph, where nodes represent genes or gene products and edges represent biochemical processes like reaction, transformation, interaction, activation, and inhibition. Similar to a GRN, a Gene Co-Expression Network (CEN) [295] is an undirected graph, where the nodes correspond to genes or gene activities and undirected edges between genes represent significant co-expression relationships. In a co-expression network, two genes are connected by an undirected edge if their activities have a significant association over a series of gene expression measurements.

Protein–Protein Interaction (PPI) Networks represent the interactions among proteins [296]. PPI networks are fundamental for cellular functions, for example, the assembly of cell structural components and the transcription process translation and active transport. The PPIs can be represented as directed or undirected graph, where the nodes represent the proteins and the edges correspond to the interaction among connected proteins. Clearly, this simple representation does not capture the aspects of interaction such as: the kind of interaction (e.g., a phosphorylation, a complexation, a colocalization), or the kind of reaction (e.g. the direction, the kinetic). If one wants to provide a distinction between a reagent and a product or one wants to represent biochemical reactions, some complexity is necessary. In this case a directed graph will represent the distinction between reagents and products. Finally, a label on the edges can specify the kind of interaction, i.e., phosphorylation, alkylation, ubiquitination. The model based on a directed graph can be similarly used to model metabolic reactions. In this case nodes can be proteins, nucleic acids, compounds, or metabolites, and edges can represent all kind of interactions. The importance of the determination of a correct model for PPI networks could be important for effective experimental planning, helping to determine possible interactions. Currently, there exists three common models used for PPI networks: Random Graph, Scale Free, and Geometric Random Graph [255]. Starting from the highest abstract level, the components of interactions (e.g., proteins and enzymes) can be modeled as a set of nodes connected by edges representing the interactions. This informal model is a graph representation according to a formal mathematical language.

13.2.1 Random graph model

The Erdos–Renyi model [297] constitutes an abstract representation of a random network in which a specified probability describes the existence of an edge between each couple of nodes.

Formally, a *random graph*, $G(n, p)$ is a graph with n nodes, where each possible edge has the probability p of existing. Consequently, the number of edges in such a graph is a random variable. $G(n, p)$ can be seen as a set of graphs with n nodes in which each graph is denoted by its probability related to its number of edges. For a random graph, the average degree z of a vertex is equal to

$$z = \frac{n(n-1)}{n} \approx np \tag{13.1}$$

for a large number of n. So, once one knows n, any property can be expressed both in terms of p or z. Consequently, this model offers the advantage of summarizing the topological properties in two parameters, n and p. In a few words, it is possible to recall that for large values of n (or alternatively when $z = 1$), random graphs exhibit a transition phase causing the formation of a so-called *giant component*. A component is a subset of nodes that are all reachable from other nodes. A giant component, consequently, is the largest component. The formation of a giant component is a characteristic of many real networks, both biological and social. Despite this, random graphs do not capture the property of the high clustering coefficient of real networks. This drawback appears also in metabolic networks as reported in [298]. In that work, the authors analyze a metabolic network of *E. coli*, by building a graph of interactions in which the vertices represent substrates and products and the edges represent interactions. The clustering coefficient of the network is 0.59, while a random graph with the same number of node possesses a value of 0.09.

13.2.2 Scale-free model

The main characteristic of scale-free networks [299] is the power-law degree distribution, that is, the probability that a generic node has exactly k edges is expressed by $P(k) = k^{\gamma}$, where γ is the degree exponent. A property of these networks is the presence of a small number of highly connected nodes (called *hubs*) that determine other properties. Generally, for these networks the clustering coefficient is independent from the number of nodes n, and the diameter is very small, following the $\log\log(n)$ formula.

13.2.3 Geometric random graph model

A geometric graph [300] $G(V, r)$ is a graph whose nodes are points in a metric space that are connected by an edge if their distance is below a threshold value r, called the radius. Formally, let $u, v \in V$, and the edge set is $E = \{\{u, v\} | (u, v \in V) \wedge (0 < \|u - v\| < r)\}$, where $\|.\|$ is a defined distance norm. Generally, a 2-D space is considered, the norms are the well-known Manhattan or Euclidean distance, and the radius takes values in (0, 1).

Thus, a random geometric graph $G(n, r)$ is a generalization of this model in which nodes correspond to n points in a metric space. Clearly, these points are distributed uniformly and independently. The properties of these graphs have been studied when $n \to \infty$ [301]. Surprisingly, certain properties of these graphs appear when a specific number of nodes is reached.

13.3 Motif discovery

In recent years considerable attention has been given to network motifs, which are specific subnetworks that appear more frequently than expected considering the degree distribution of a biological network. The interest in discovering motifs in networks relies on two main aspects that concern the individuation of small subnetworks that play important roles, and the unraveling of the evolutionary mechanism in biological networks. In fact, motif analysis has revealed the existence of subsequences that plays important biological roles, for example, protein binding zones or particular RNA-coding substrings. In protein–protein interaction networks, motif detection represents a computational challenge. The reason relies on the consideration that the network motif in a protein interaction network can be similarly defined as a pattern of interconnections recurring more frequently than expected by chance, where a pattern of interconnection is represented by a subgraph.

Thus, considered from the computational point of view, the number of possible subgraphs grows exponentially in a large network. On the other hand, considered from a biological point of view, a pattern is considered statistically significant when it is determined by following an evolutionary mechanism. In general, the motifs can be classified according to the network model. For example, in undirected networks, the size and the structure of the induced subgraphs may be used as discriminating parameters. So, it is possible to distinguish:

- paths of a given size, i.e., the number of edges in the path;
- cliques of size k, i.e., the complete subgraph of a graph;
- loops of size k, i.e., the edges that connect a vertex to itself;
- stars, i.e., a special type of subgraph in which $n-1$ nodes have degree 1 and a single node have degree $n-1$;
- pentagons, i.e., subgraphs in which the union of its nodes forms a pentagon.

The motif discovery can be conducted by enumerating all the subgraphs subdivided according to the number of nodes and then by searching for the statistical significance. In particular, the detection of motifs in a network requires initially finding subgraphs that occur in the input graph and in their quantity. After that, it is possible to determine which of these subgraphs are topologically equivalent; then, they are grouped into different classes. Finally, the subgraph classes appearing at a much higher frequency than in random graphs are displayed.

The different typology of motifs is represented by colored motifs. The ones that are defined as multisets of nodes labels, i.e., colors. Thus, the occurrence of a motif consists of connected subgraphs in which the node labels match the motif.

13.4 Network embedding (representation learning)

Recently, there is growing interest to introduce novel approaches for encoding structural information about the graph, also known as *representation learning* [302,303]. The idea behind these approaches is to learn a mapping for nodes (or subgraphs) as points of a low-dimensional vector space $\mathbb{R}^d$ [304]. The goal of such methods is to realize a mapping so that geometric relationships among embedded objects reflect the structure of the original graph. Afterwards, the obtained embeddings may be used in other machine learning tasks (e.g., node classification) or in other graph analysis algorithms.

Despite the relatively short time since such approaches have been introduced, there exists many algorithms and many classification attempts described in some existing surveys [254,260,264,304–308]. Here, we follow the classification proposed by Hamilton et al. by presenting a general overview and a discussion of some related methods, while the interested reader may refer to such works for deeper discussions.

In the rest of the section, we will assume that the input of representation learning algorithms is a undirected and unweighted graph $G=(V,E)$ with its associated adjacency matrix A and a real-valued matrix X containing node attributes $X \in R^{mx|V|}$. The goal of each algorithm is to map each node into a vector $z \in \mathbb{R}^d$, where $d < |V|$.

As introduced by Hamilton et al., all the embedding algorithms may be represented using two mapping functions: an encoder ($Enc : V \rightarrow R^d$), that maps each node to the embedding and a decoder Dec that decodes structural information about the graph from the learned embeddings. In parallel, it is possible to introduce one more function: a pairwise similarity function that quantifies the similarity among nodes (or substructure) of the graph G and a loss function that measures the difference between the similarity of two nodes and the similarity among their embeddings $l(v_i, v_j), (e_{v_i}, e_{v_j})$. The simplest similarity function is the adjacency matrix that accords similarity equal to 1 to adjacent nodes and equal to 0 elsewhere, while a common similarity function among embeddings is the dot product between the embedding vectors.

The desired behavior of an embedding algorithm is that, given two nodes v_i and v_j, the similarity among the embeddings should reflect the similarity among nodes, i.e., minimizing the loss l. Formally, given two nodes v_i, and v_j in V, their embeddings e_{v_i} and e_{v_j}, and the loss among the similarity of nodes and of the embeddings l, the embedding algorithm should train its parameters to minimize the overall loss function $L = \sum_{v_i, v_j \in V} l(v_i, v_j), (e_{v_i}, e_{v_j})$.

Shallow embedding methods

The first embedding methods rely on a simple embedding defined by Hamilton et al. as *shallow embedding*. In these methods each node is encoded into a vector using a simple function such as

$$ENC(v_i) = M v_i, \tag{13.2}$$

where M is a matrix containing the embedding vectors and v_i is a vector used for selecting the column. The matrix M contains all the embeddings [309]. Each column corresponds to a node, and the number of rows d is lower than the number of nodes n. These embeddings were initially inspired by matrix factorization approaches. The differences between these methods reside in the use of different loss functions and similarity measures. Examples of such methods are GraRep [310] and Hope [311].

The main drawback of these methods is that they consider as a similarity measure only the presence of edges among nodes, i.e., they consider only the first-order neighborhood of nodes. Conversely, nodes may be structurally similar even if they are far distant within the networks.

Random walk-based methods

Consequently, a set of methods investigating higher-order neighborhoods (or local structures) have been introduced. A set of effective methods that have been introduced is based on random walks. A random walk is a random path that consists of a succession of random steps over the networks. The key idea of these methods is to derive the similarity among two nodes on the basis of the co-occurrence of nodes within random walks, i.e., two similar nodes have a greater chance to co-occur in random walks. In this way, the higher-order neighborhood between two nodes is encoded by the use of random walks.

The overall strategy of these methods may be summarized in three main points:

1. Simulation of random walks starting from each node;
2. For each node v, store the sequence of nodes visited by random walks starting from v;
3. F for each node v, learn its embedding starting from the sequence just described.

The difference between these methods resides in the particular definition of random walks. For instance, **DeepWalk** uses fixed-length, unbiased random walks starting from each node [312]. Node2Vec, instead, uses random walks having different lengths, defined as biased random walks, [313]. Ribeiro et al. proposed struct2vec that belongs to the same class of node2vec. In struct2vec, biased random walks are generated and propagated in a modified version of the network. In this network, the nodes, that are structurally similar in the original network, are close in distance. struct2vec derives representations for nodes in four steps:

- The input network is analyzed to determine the structural similarity between each node pair, and the similarity is used to derive a hierarchy of similarity;
- The hierarchy is used to build a weighted multilayer graph in which each layer corresponds to a level of similarity;
- Run biased random walks within the weighted multilayer graph;
- Derive a representation of the nodes from the sequence obtained by random walks.

Similarly, **LINE** [314] combines first- and second-order proximity to take into account both the local and global network structures. It is based on an edge-sampling algorithm to address the limitations of the classical stochastic gradient descent. The first-order proximity is evaluated considering the pairwise proximity between any pair of nodes, while the second-order proximity is evaluated on the basis of the neighboring nodes.

13.5 Networks alignment

The reconstruction of the evolution and of its impact on the organisms is a major goal of biology. The advent of the technologies investigating the complete organism at the molecular level has launched the challenge to reconstruct also the evolutionary changes at the molecular level. In other words, researchers have proposed the hypothesis that the record of evolutionary changes is present at the genomic level. Consequently, if two organisms are descendant from a single ancestor, their DNA will be similar. To confirm this hypothesis, a number of algorithms for the comparison of the nucleic acids have been successfully introduced, like BLAST [315]. The alignment algorithms compare the sequences of DNA of two (pairwise alignment) or more (multiple alignment) species with respect to a null model. The null model is used to discard similarities between sequences that are due to chance and are not due to evolution. The same conceptual model has been introduced in the interaction networks in which the counterpart of sequence alignment yield to the introduction of protein-interaction network alignment. Such a field of research aims to elucidate the conservation among evolution of molecular machineries or, similarly, the conservation of protein complexes among species. Considering the representation of interaction networks, such analysis can be formulated as the search for similar or equal subgraphs in two or more networks [316]. This formulation permits the definition of a graph alignment as a counterpart of sequence alignment, based on an appropriate scoring function that measures the similarities of two subgraphs and that enables the global comparison of interaction networks by the measurement of mutually similar subgraphs. From an algorithmic perspective, this problem is not simple because all its formulations can be translated into a graph matching. In other words, the graph matching entails finding a mapping one-to-one from a node in a graph to a node in the another graph. Such a mapping starts with the choice of a few numbers of proteins (also referred as seed proteins), and successively it grows the alignment graph [317] by analyzing the graph topology. The seed proteins are selected in order to manage the (sub)-graph isomorphism problem related to the alignment problem, which is computationally hard in some general formulations. Considering the interaction networks, the mapping from one species to another resides in the definition of such a mapping, i.e., how to state that two proteins of two different species can be associated [318]. The correct choice of the association, or alternatively the choice of a similarity scoring function among nodes, is a key problem [319,320]. Usually, the associated proteins should be orthologs, in particular, when the alignment is used to explain evolutionary relationship. The definition of orthologs is often an excessively restrictive criterion, and, for this reason, the different algorithms try to choose correspondent proteins on the basis of their sequence similarity, and eventually by introducing other biological considerations. Starting from the seed proteins, the graph alignment algorithms build an alignment allowing also inexact matching, i.e., allowing gaps, mismatches, node deletions, and insertions. A key point in the building of the alignment is the topology of the substructure. While, initially, algorithms tried to evidence shared linear paths, a number of algorithms evidence common dense subgraphs (i.e., protein complexes), and some approaches allowing the search for arbitrary topologies have been introduced. Network alignment (NA) is a computational technique widely used for comparative analysis of PPI networks between species, in order to predict evolutionary conserved components or sub-structures on a system data level. Network alignment is a common problem that requires searching for a node mapping that best fits one network into another network. The problem of graph alignment consists of the mapping between two or more graphs to maximize an associated cost function that represents the similarity among nodes or edges. Formally, let $G_1 = \{V_1, E_1\}$ and $G_2 = \{V_2, E_2\}$ be two graphs, where $V_{1,2}$ are the sets of nodes and $E_{1,2}$ are the sets of edges; the **graph alignment problem** consists of finding an alignment function (or a mapping) $f : V_1 \rightarrow V_2$ such that the similarity between mapped entities is maximized. Thus, the alignment problem relies on the *(sub)-graph isomorphism problem*, which is computationally NP-hard in some general formulations [321], and it should be resolved by heuristic methods. The similarity between the graphs is defined by a cost function, $Q(G_1, G_2, f)$, also known as the quality of the alignment. Q expresses the similarity between two input graphs on a specific alignment f, and the formulation of Q strongly influences the mapping strategy [322]. In the literature, there exist various network alignment algorithms that investigate approximate solutions. The network alignment algorithms can be categorized according to the kind of input, pairwise or multiple alignment, and the scope of node mapping desired, i.e., local or global alignment.

13.5.1 Global network alignment

Global Network Alignment (GNA) algorithms search for the best mapping that cover all nodes of the input networks by generating a one-to-one node mapping. This strategy takes into account only the topology of input networks, leaving out the similarity among small regions (representing conserved motifs). In general, GNA exploits a two-step schema to build the alignment. At first, the algorithms adopt a cost function that maximizes the node similarity (or node conservation) or the number of conserved edges (edge conservation) to estimate the similarity between pairs of nodes. Then, they apply an alignment method to find a high-scoring alignment on the basis of total similarity over all aligned nodes among all possible alignments.

An example of a GNA algorithm is MAGNA [322], which applies a genetic algorithm to build an improved alignment. MAGNA simulates a population of alignments, and then it selects the best one. MAGNA++ [323] extends MAGNA enabling both the maximization of any different measures of edge conservation and any desired node conservation measure. Also, SANA (Simulated Annealing Network Aligner) [324] uses simulated annealing to build a final alignment. It takes as input two networks and an initial alignment generated randomly or chosen using other aligners. More recently, Malod–Dognin et al. [325] presented UAlign, which unifies global alignments built by different network aligners. The goal of the authors was to overcome the limitations of global network aligners in the coverage of built alignments. Others GNAs are IsoRank [326], GRAAL [327] and the GRAAL family (H-GRAAL [328], MI-GRAAL [329], C-GRAAL [330], L-GRAAL [331]), GHOST [332], WAVE [333], and IGLOO [334].

13.5.2 Local network alignment

Local Network Alignment (LNA) algorithms search multiple small subnetworks with high similarity among input networks by producing a many-to-many node mapping. These subnetworks are conserved patterns of interaction that can correspond to conserved motif or pattern of activities. LNA algorithms employ a two-step schema to build the alignment. At first, starting from a list of *seed nodes* selected from biological considerations, the algorithms integrate all the information in an auxiliary structure usually referred to as alignment graph. Then they mine the graph to evidence interesting regions. An example of a GNA algorithm is AlignNemo [317], which enables the discovery of subnetworks of related proteins in biological function and topology. The algorithm can deal even with sparse interaction networks by analyzing the topology of nodes adjacent to the proteins directly interacting with the current solution. AlignMCL [281] is an evolution of the previous AlignNemo. AlignMCL builds the local alignment by merging all the input data into a single graph (the alignment graph) that is afterwards clustered by using the Markov cluster algorithm MCL [335] to mine the conserved subnetworks. Also, GLAlign (Global Local Aligner) is a novel local network alignment methodology [336] that applies a global network aligner to produce an initial seed of nodes that are selected using the topological information only. Then, GLAlign combines the topology information from global alignment with biological information (e.g., homology relationships) by applying a linear combination of weights that combines topological and biological information. Finally, it uses this global mapping as input for a local network aligner. Others LNA algorithms are NetworkBLAST [337] and NetAligner [338].

13.5.3 Pairwise and multiple alignment

Another way to formulate network alignment problem considers the number of networks that need to be aligned: *pairwise or multiple alignment*. The pairwise network alignment (PNA) takes as input two networks with the goal to detect similar regions in two networks. The multiple network alignment (MNA) builds the alignment among three or more networks with the goal to find aligned node clusters. Pairwise and multiple network alignment can be local or global, with node mappings one-to-one or many-to-many.

The pairwise alignments can search the similar small subnetworks exploiting many-to-many mapping between nodes of the compared network or can look for the best overlap of the whole compared networks exploiting one-to-one node mapping. In an alignment of multiple networks, a many-to-many MNA occurs when an aligned cluster contains more than one node from a single network, whereas, when one node per network is present in every aligned cluster, one-to-one MNA occurs. In the literature, both PNA and MNA are applied to build the alignment of protein interaction networks [285]. However, it has been discovered that MNA leads to deeper biological information than PNA because it is able to capture functional knowledge that is common to multiple species.

There exist various proposed MNA algorithms in the literature. For example, MultiMAGNA++ [339] is a global one-to-one MNA algorithm that uses a genetic algorithm to build an improved alignment directly optimizing both edge and node conservation, while the alignment is being constructed. Specifically, MultiMAGNA++ simulates a population of alignments that evolves over time by applying the genetic algorithm and a new function for the crossover of parent alignments into a superior child alignment that allows for aligning multiple networks. Other algorithms are GEDEVO-M [340], LocalAli [341], IsoRankN [326], SMETANA [342], FUSE [343], NetCoffee [344], and BEAMS [345].

Fig. 13.1 summarizes the various alignment types.

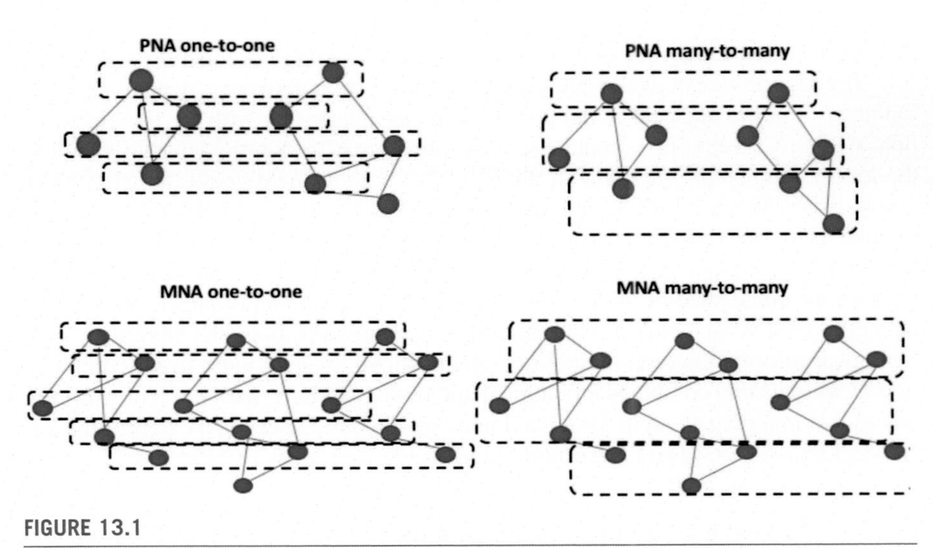

FIGURE 13.1

An example of PNA one-to-one, PNA many-to-many, MNA one-to-one, and MNA many-to-many.

13.5.4 Quality evaluation of network alignment

The evaluation of pairwise network alignment algorithms can be performed by considering both topological and biological aspects (see [346] for a complete discussion).

The topological quality is related to alignment algorithm capability to reconstruct true node mapping and to conserve much possible edges.

A widely used measure is *node correctness (NC)* [327] that is defined as the fraction of nodes of one network that are correctly mapped to nodes of the other networks concerning the true node mapping. NC is not used for local network alignments since some local network alignment algorithms may map a node from a network to many nodes of the other network [347]. Meng et al. [334] defined three novel measures: Precision *P-NC*, Recall *R-NC*, and F-score *F-NC* that may be used for both global and local alignments. Let suppose that the alignment f produces a set of node pairs composed of N_{al} nodes, while the true node mapping is composed of M_{tr} nodes. *P-NC* is calculated as $\frac{M_{tr} \cap N_{al}}{M_{tr}}$. *R-NC* is defined as $\frac{M_{tr} \cap N_{al}}{N_{al}}$. *F-NC* is the harmonic mean of the two previous measures.

Similarly, to measure how many edges are correctly mapped in an alignment, three popular measures have been proposed: edge correctness (EC) [327], induced conserved structure (ICS) [332], and symmetric substructure score (S3) [322], which outperforms the previous ones. Similarly to node correctness, the S3 cannot be used directly to evaluate the quality of local network alignment algorithms. Therefore, other measures have been defined [347], such as generalized S3 (GS3) and high-node coverage S3 (NCV-S3).

The evaluation of alignment biological quality is performed by applying the following measures: *Gene Ontology (GO) Correctness* that identifies the number of aligned protein pairs that are annotated with the same GO terms and *Precision (P-PF)*, *Recall (R-PF)*, and *F-score (F-PF)* that measure the accuracy of known protein-function predictions. In particular, P-PF is the fraction of the union of the set of predicted protein-GO term associations and the set of true protein-GO term associations out the set of predicted protein-GO term associations. R-PF is the fraction of the union of the set of predicted protein-GO term associations and the set of true protein-GO term associations out the set of true protein-GO term associations. F-PF is the harmonic mean of P-PF and R-PF.

Also, a topological and functional quality assessment is performed for multiple alignments. The adopted measures to evaluate the topological alignment quality are: *Adjusted node correctness (NCV-MNC)*, *Adjusted cluster interaction quality (NCV-CIQ)*, and *Largest common connected subgraph (LCCS)*. NCV-MNC is the geometric mean of the node coverage (NCV), defined as nodes forming an alignment cluster with respect to all nodes in the networks; cluster consistency (MNC) is defined as 1 minus the mean of normalized entropy (NE) of all clusters in the alignment. NCV-CIQ is the geometric mean of the node coverage (NCV) and CIQ that measures edge conservation, and it is the generalization of the established S3. LCCS is the geometric mean of the fraction of number of nodes in the largest common connected subgraph out the maximum possible number of nodes in the one and the fraction of number of edges in the largest common connected subgraph out the maximum possible number of edges in the other. To evaluate the functional alignment quality for MNA, three measures are defined: *Mean Normalized Entropy (MNE)*, *GO correctness (GC)*, and *Accuracy of protein function prediction*. MNE measures the internal cluster consistency, and it defined as the mean of the normalized entropy (NE) across all clusters in the alignment. In the case of multi-alignment, the NE takes into account the number of unique GO terms, the number of proteins annotated with GO term, and the total number of protein-GO term annotations in the cluster. GC is the fraction of protein pairs that share at one or more GO terms. The accuracy of protein-function prediction is computed according to Precision (P-PF), Recall (R-PF), and F-score (F-PF), as previously defined. Table 13.1 summarizes the various network alignment algorithms.

Table 13.1 Network alignment algorithms.

Algorithm	Node Mapping Scope	Input Kind	Node Mapping	Evaluation Measures
GRAAL	GNA	PNA	One-to-one	P-NC, R-NC, F-NC, GS3, NCV-S3, GO correctness , P-PF, R-PF, F-PF
H-GRAAL	GNA	PNA	One-to-one	P-NC, R-NC, F-NC, GS3, NCV-S3, GO correctness , P-PF, R-PF, F-PF
MI-GRAAL	GNA	PNA	One-to-one	P-NC, R-NC, F-NC, GS3, NCV-S3, GO correctness , P-PF, R-PF, F-PF
C-GRAAL	GNA	PNA	One-to-one	P-NC, R-NC, F-NC, GS3, NCV-S3, GO correctness , P-PF, R-PF, F-PF
L-GRAAL	GNA	PNA	One-to-one	P-NC, R-NC, F-NC, GS3, NCV-S3, GO correctness , P-PF, R-PF, F-PF
IsoRank	GNA	PNA	One-to-one	P-NC, R-NC, F-NC, GS3, NCV-S3, GO correctness , P-PF, R-PF, F-PF
GHOST	GNA	PNA	One-to-one	P-NC, R-NC, F-NC, GS3, NCV-S3, GO correctness , P-PF, R-PF, F-PF
WAVE	GNA	PNA	One-to-one	P-NC, R-NC, F-NC, GS3, NCV-S3, GO correctness , P-PF, R-PF, F-PF
MAGNA	GNA	PNA	One-to-one	P-NC, R-NC, F-NC, GS3, NCV-S3, GO correctness , P-PF, R-PF, F-PF
MAGNA++	GNA	PNA	One-to-one	P-NC, R-NC, F-NC, GS3, NCV-S3, GO correctness , P-PF, R-PF, F-PF
SANA	GNA	PNA	One-to-one	P-NC, R-NC, F-NC, GS3, NCV-S3, GO correctness , P-PF, R-PF, F-PF
IGLOO	GNA	PNA	One-to-one	P-NC, R-NC, F-NC, GS3, NCV-S3, GO correctness , P-PF, R-PF, F-PF
NetworkBLAST	LNA	PNA	Many-to-many	P-NC, R-NC, F-NC, GS3, NCV-S3, GO correctness , P-PF, R-PF, F-PF
NetAligner	LNA	PNA	Many-to-many	P-NC, R-NC, F-NC, GS3, NCV-S3, GO correctness , P-PF, R-PF, F-PF
AlignNemo	LNA	PNA	Many-to-many	P-NC, R-NC, F-NC, GS3, NCV-S3, GO correctness , P-PF, R-PF, F-PF
AlignMCL	LNA	PNA	Many-to-many	P-NC, R-NC, F-NC, GS3, NCV-S3, GO correctness , P-PF, R-PF, F-PF
LocalAli	LNA	PNA	Many-to-many	P-NC, R-NC, F-NC, GS3, NCV-S3, GO correctness , P-PF, R-PF, F-PF
GLAlign	LNA	PNA	Many-to-many	P-NC, R-NC, F-NC, GS3, NCV-S3, GO correctness , P-PF, R-PF, F-PF
MultiMAGNA++	GNA	MNA	One-to-one	NCV-MNC, NCV-CIQ, LCCS, MNE, GC, P-PF, R-PF, F-PF
GEDEVO-M	GNA	MNA	One-to-one	NCV-MNC, NCV-CIQ, LCCS, MNE, GC, P-PF, R-PF, F-PF
IsoRankN	GNA	MNA	Many-to-many	NCV-MNC, NCV-CIQ, LCCS, MNE, GC, P-PF, R-PF, F-PF
SMETANA	GNA	MNA	Many-to-many	NCV-MNC, NCV-CIQ, LCCS, MNE, GC, P-PF, R-PF, F-PF
LocalAli	LNA	MNA	Many-to-many	NCV-MNC, NCV-CIQ, LCCS, MNE, GC, P-PF, R-PF, F-PF
NetCoffee	GNA	MNA	Many-to-many	NCV-MNC, NCV-CIQ, LCCS, MNE, GC, P-PF, R-PF, F-PF
BEAMS	GNA	MNA	Many-to-many	NCV-MNC, NCV-CIQ, LCCS, MNE, GC, P-PF, R-PF, F-PF

CHAPTER

Biological pathway analysis

14

CONTENTS

Abstract

In the analysis of a list of differential enriched genes obtained using microarrays, Genome Wide Association Studies (GWAS) and Pathways Enrichment Analysis (PEA) play a central role in elucidating complicated gene interactions involved in complex human diseases. Pathway enrichment analysis provides a procedure that enables a comprehensive understanding of complex diseases' molecular mechanisms. The general principles and methods for the pathway enrichment analysis are introduced. Then, a general and accurate review describes some of the well-known pathway databases, and some software tools used to perform pathway enrichment analysis.

Artificial Intelligence in Bioinformatics. https://doi.org/10.1016/B978-0-12-822952-1.00024-3

Keywords

DEGs, GWAS, Microarray, Statistical analysis, Pathway enrichment analysis, Biological pathways, Complex disease

14.1 Introduction

The available omics experimental investigation methodologies (genomics, metabolomics, proteomics, and transcriptomics), i.e., high-throughput methods (HT), produce a massive amount of data for a single experiment, leading to the exponential growth of omics data that can be sequenced and stored in large quantities for medical research and advanced genomic medicine. Gene expression profiling based on DNA microarrays is a widespread method employed to identify the genes responsible for a specific condition in case-control studies. The obtained list of differentially expressed genes (DEGs) could be used to define better therapies for complex diseases, like cancer and Parkinson's, or to develop additional effective drugs. The principal obstacle in adopting DEGs' lists in clinical activities or drug development is the missing information about the molecular mechanisms altered by the DEGs.

In addition to HT methods, the continuous improvement in the available laboratory methods employed to identify biochemical reactions has contributed to identifying many previously unknown molecular mechanisms. Biochemical reactions are essential in figuring out how cellular machinery works. Biochemical reactions are also known as signaling cascades, a sequence of chemical effects inside a cell triggered by a specific signal. Biochemical cascades are a series of events in which one event triggers the next linearly. At each step of the signaling cascade, various controlling factors regulate cellular actions to respond adequately to cues about their changes induced by their internal and external environments, e.g., activating transcriptional factors, regulating gene expression and activating or inhibiting targeted proteins.

In particular, the cellular machinery relies on the interactions of several types of molecules, including large molecules like proteins, DNA, RNA, and small molecules, such as lipids, hormones, and metabolites essential to accomplish their tasks. Thus, there is a growing need to represent the interconnections among genes, proteins, and large and small molecules within a living organism that are responsible for triggering the biochemical reactions cascade activating or inhibiting specific biological functions. The signaling cascade can be represented as an interaction network where the genes, proteins, and the molecules are the network's nodes. At the same time, the connections among the nodes (i.e., the interactions) are the network's edges.

Biological networks can be represented using graph theory. Formally, graphs comprise a discrete mathematical object defined as a couple $G =< V, E >$, where $V = v_1, v_2, \ldots v_n$ is the set of nodes; whereas, $E =< V \times V >$ is the set of edges connecting interacting nodes among them. The vertices of a graph are sometimes called nodes or dots; edges are sometimes called links, lines, arcs, or connections.

The set of vertices of ***G*** is denoted by ***V(G)***, and the set of edges is denoted by ***E(G)***. In short form, a graph can be referred to as *G*.

14.2 Biological pathways

Biological pathways are used to represent the interaction networks that happen within the cellular machinery. Biological Pathways (BPs) are schematic overviews of the biochemical reaction cascades that occur inside cells in numerous living organisms. Pathways control almost the totality of biological events of an organism, and it is better to describe them as classes rather than individually.

Pathways can be divided into these main classes: *Metabolic*, *Gene Regulation*, and *Signal Transduction*.

- *Metabolic Pathways* are the pathways that regulate the chemical reactions carried out by a cell, for example, to transform food into energy. Fig. 14.1 is a graphical representation of a metabolic pathway.

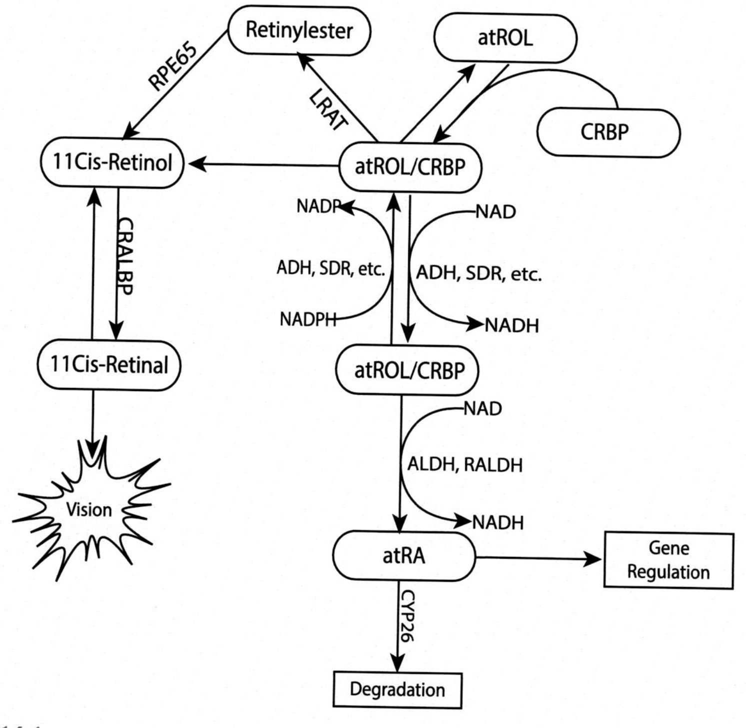

FIGURE 14.1

Graphical representation of a generic metabolic pathway.

- *Gene Regulation Pathways* consist of pathways responsible for genes activation or inhibition. Gene regulation is vital for every organism because genes contain critical information for all the cell processes, including protein production. Fig. 14.2 is a graphical representation of a gene regulatory pathway.

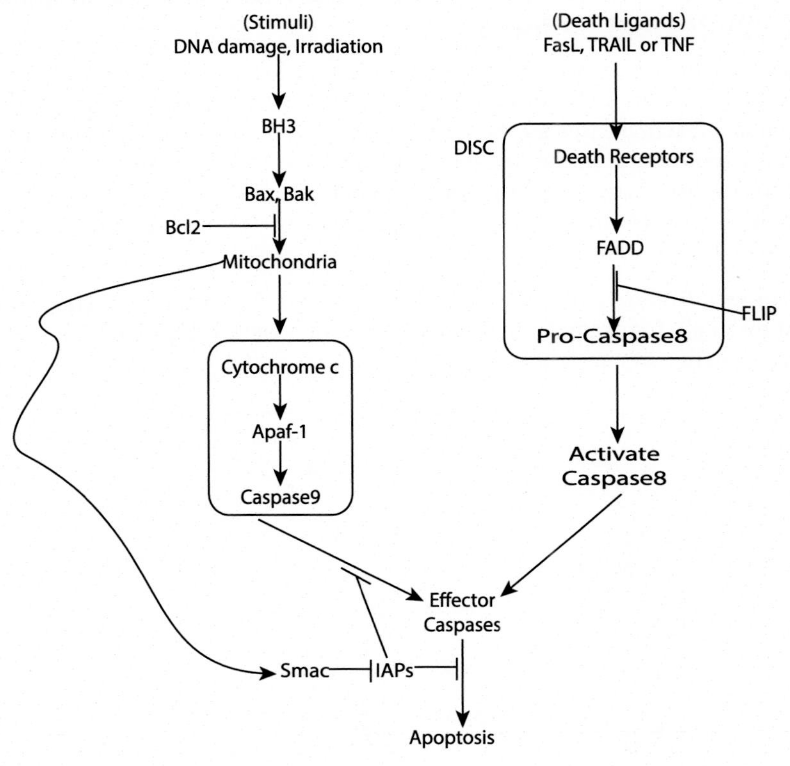

FIGURE 14.2

Graphical representation of a generic gene regulatory pathway.

- *Signal Transduction Pathways* comprise the pathways that can deliver a signal from a source to a destination. They transmit a signal from a cell to the target receptor located on the cell membrane. Fig. 14.3 graphically represents a signal transduction pathway.

Hence, it is possible to use pathway enrichment methods to discover the connection between a DEGs' list and the involved biological mechanisms underlying complex diseases. Pathways have become the first choice for gaining insight into the underlying biology of differentially expressed genes and proteins because they reduce complexity and offer increased explanatory power. Pathway enrichment regarding a list of genes and their interactions (gene regulations, protein–protein interactions, metabolic reactions, etc.), usually obtained from public and private biological

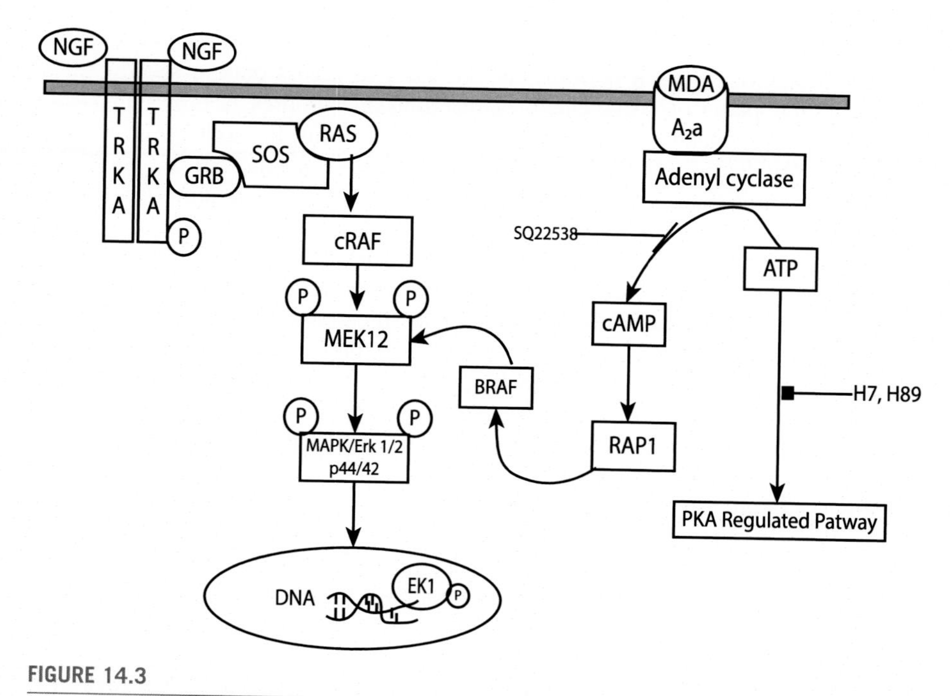

FIGURE 14.3

Graphical representation of a generic signal transduction pathway.

pathway databases, serve as repositories of current knowledge pathways. Pathway databases comprise KEGG, REACTOME, and BioCyc, to name some examples.

14.3 Pathway databases

Pathway databases are increasing in number in recent years, providing biologists with many data sources to collect, integrate, and analyze information to support their activities research. Here, we describe some of the well-known pathway databases.

- Kyoto Encyclopedia of Genes and Genomes (KEGG) [348] is a multispecies database, including genomic, chemical, pathway, and phenotypic information. KEGG consists of 18 original databases in four categories, and it collects pathways in classes, including metabolic pathways, genetic information pathways, and signaling pathways from several organisms, including humans. Pathways are manually curated by the experts in the literature. KEGG-API or KEGG-FTP allow users to download pathways in KGML format, the KEGG XML format used to encode the pathways. KEGG also provides a web interface to browse each pathway.

- MetaCyc [349] is a curated database of experimentally illustrated metabolic pathways. MetaCyc makes it possible for the users to access data in several ways by searching pathways using enzymes, reactions, metabolites, or browsing MetaCyc pathways. MetaCyc data can be downloaded using the following formats: BioPAX, Pathway Tools attribute-value, Pathway Tools tabular, SBML, and Gene Ontology annotations. To access and download data, MetaCyc requires a paid subscription.
- Protein Analysis Through Evolutionary Relationships Classification System (Panther) [350] is a web resource intended to classify proteins (and their genes) promoting high-throughput analysis. Panther Classifications are the result of human curation through the literature. Panther includes protein sequencing, evolutionary information, and metabolic and signaling pathways information. In its current version, Panther stores pathways from several organisms, including humans. Pathway data can be accessed using the Panther API or downloaded in the BioPAX, SBML, and Protein Sequence Association data format.
- PathwayCommons [351] is an open-access portal to a collection of public pathway databases such as Reactome and Cancer Cell Map, as well as protein–protein interaction databases, such as HPRD and IntAct. PathwayCommons provides technology for integrating pathway information, a web interface to browse pathways, and a web service API for automatic access to the data. Pathway data can be downloaded in PSI-MI and BioPAX formats.
- Reactome [352,353] is an open-source manually curated and peer-reviewed database containing biological processes, biochemical reactions, and pathways, from several species including humans. In the current version, Reactome includes the whole known pathways of humans and from other organisms. Reactome enables browsing pathways through the graphical web interface and downloading the data in various formats comprising SBML Level 2 and BioPAX Level 2 and Level 3.
- WikiPathways [354] is an open pathway database devoted to improve the ongoing efforts, such as KEGG, Reactome, and PathwayCommons databases. WikiPathways has a pathway visualization web page presenting pathways diagrams, download options, gene, and protein lists. In addition, any pathway can be edited from the embedded pathway editor. WikiPathways data are freely available for download as images and in GPML, a custom XML format, or programmatically using the available Webservice or API.

14.4 Pathway representation formats

The number of pathway databases has been increasing in recent years, making it possible to support researchers in their activities. Often, biologists need to use information from many sources to support their research. But, on the other hand, each database adopted its own representation formats, making data integration from multiple databases challenging. Thus, there is the need to define a unique file format that makes it possible to integrate data from various databases. Pathway data are encoded using various types of formats such as: *BIOPAX-LEVEL 1,2,3*, *PSI-MI*, *SBML*, and *CellML*.

The high level of heterogeneity among available pathway databases contributes to limiting the effectiveness of automatic data analysis.

14.4.1 BioPAX

BioPAX (Biological Pathway Exchange) [355] is a metalanguage defined in *OWL DL* and is coded in the *RDF/XML* format. BioPAX aims to facilitate the integration and exchange of data maintained in biological pathway databases through a unique way to represent data. Traditionally, integrating and comparing various different data from a syntactic and semantic perspective has always been a challenge in bioinformatics. The latest version is BioPAX Level 3, and it is the recommended version. Because of that, BioPAX Level 3 supports metabolic pathways, signaling pathways (including states of molecules and generic molecules), gene regulatory networks, molecular interactions, and genetic interactions. It is not entirely backward compatible with Levels 1 and 2. Fig. 14.4 conveys a biological pathway encoded in the BioPAX Level 3 format.

```
<?xml version="1.0" encoding="UTF-8"?>
<rdf:RDF xmlns:rdf="http://www.w3.org/1999/02/22-rdf-syntax-ns#"
xmlns:bp="http://www.biopax.org/release/biopax-level3.owl#"
xmlns:rdfs="http://www.w3.org/2000/01/rdf-schema#"
xmlns:owl="http://www.w3.org/2002/07/owl#"
xmlns:xsd="http://www.w3.org/2001/XMLSchema#"
xml:base="http://www.reactome.org/biopax/56/176806#">
  <owl:Ontology rdf:about="">
    <owl:imports rdf:resource="http://www.biopax.org/release/biopax-level3.owl" />
  </owl:Ontology>
  <bp:Pathway rdf:ID="Pathway1">
    <bp:pathwayComponent rdf:resource="#Pathway2" />
    <bp:pathwayComponent rdf:resource="#Pathway6" />
    <bp:pathwayComponent rdf:resource="#Pathway7" />
    <bp:pathwayComponent rdf:resource="#Pathway12" />
    <bp:pathwayOrder rdf:resource="#PathwayStep17" />
    <bp:pathwayOrder rdf:resource="#PathwayStep50" />
    <bp:pathwayOrder rdf:resource="#PathwayStep42" />
    <bp:pathwayOrder rdf:resource="#PathwayStep23" />
    <bp:organism rdf:resource="#BioSource2" />
    <bp:displayName rdf:datatype="http://www.w3.org/2001/XMLSchema#string">Mycobacterium
    tuberculosis biological processes</bp:displayName>
    <bp:xref rdf:resource="#UnificationXref517" />
    <bp:xref rdf:resource="#UnificationXref518" />
    <bp:comment rdf:datatype="http://www.w3.org/2001/XMLSchema#string"> ... </bp:comment>
    <bp:xref rdf:resource="#PublicationXref51" />
    <bp:xref rdf:resource="#PublicationXref52" />
    <bp:dataSource rdf:resource="#Provenance1" />
    <bp:comment rdf:datatype="http://www.w3.org/2001/XMLSchema#string"> ... </bp:comment>
    <bp:comment rdf:datatype="http://www.w3.org/2001/XMLSchema#string"> ... </bp:comment>
    <bp:comment rdf:datatype="http://www.w3.org/2001/XMLSchema#string"> ... </bp:comment>
  </bp:Pathway>
```

FIGURE 14.4

BioPAX Level 3 biological pathway encoding.

14.4.2 CellML

CellML [356] is an XML-based markup language for describing mathematical models. The purpose of CellML is to store and exchange computer-based mathematical models. CellML enables scientists to share models even if they are using different model-building software. It also enables them to reuse components from one model in another one, thus accelerating model building. CellML allows the description of biological models, including information about the model structure, equations describing the underlying processes, and additional information, simplifying the search for specific models or model components in a database. The CellML goal is to provide a comprehensive wordlist for describing biological information at various resolutions from the subcellular to the organism level. Fig. 14.5 shows a pathway encoded using the CellML format.

14.4.3 HUPO

HUPO PSI Molecular Interactions Tab-Delimited format (PSI-MITAB) [357] is a tabular data-exchange format suitable to describe interactions among biological entities. PSI-MITAB improves the annotation and representation of molecular interaction data and the accessibility of molecular interaction data for the user community. As a result, data can be downloaded from multiple sources and easily integrated. Fig. 14.6 provides an example of PSI-MITAB pathway encoding.

14.4.4 SBML

Systems Biology Markup Language (SBML)[1] is an XML-based language for representing and exchanging models of biological systems of different complexity. SBML is a modular language where each level offers a collection of components and features the fundamentals for describing a wide range of dynamical models, simulated using ordinary differential equations (ODEs) and stochastic simulations. The ongoing development of SBML contributed to enhancing its extensibility, portability, and reusability, making it the most suitable format for describing whole-cell models. SBML levels are intended to coexist: The definition of a new level does not make the previous one obsolete. Usually, a novel level introduces new functionalities, constructs, and features, enhancing analyses and simulations performance. Fig. 14.7 provides an example of SBML pathway encoding.

14.5 Pathways enrichment analysis methods

Approaches for pathway enrichment analysis can be broadly divided into three different types. In the following the three pathway enrichment analysis methods are described.

[1] https://sbml.org.

```
<?xml version='1.0' encoding='utf-8'?>
<model xmlns="http://www.cellml.org/cellml/1.0#"
xmlns:cmeta="http://www.cellml.org/metadata/1.0#"
xmlns:dc="http://purl.org/dc/elements/1.1/"
xmlns:rdf="http://www.w3.org/1999/02/22-rdf-syntax-ns#"
xmlns:bqs="http://www.cellml.org/bqs/1.0#"
xmlns:cellml="http://www.cellml.org/cellml/1.0#"
xmlns:dcterms="http://purl.org/dc/terms/"
xmlns:vCard="http://www.w3.org/2001/vcard-rdf/3.0#"
cmeta:id="asthagiri_model_2001" name="asthagiri_model_2001">
<documentation xmlns="http://cellml.org/tmp-documentation">
<article> ... </article>
</documentation>
  <units name="minute"> ... </units>
  <units base_units="yes" name="number"/>
  <units base_units="yes" name="cell"/>
  <units name="number_per_cell"> ... </units>
  <units name="molar"> ... </units>
  <units name="first_order_rate_constant"> ... </units>
  <units name="second_order_rate_constant"> ... </units>

  <component name="environment">
    <variable units="minute" public_interface="out" name="time"/>
  </component>

  <component name="c" cmeta:id="c">
    <variable units="molar" public_interface="out" name="c" initial_value="0.0"/>
    <variable units="molar" public_interface="in" name="c2"/>
    <variable units="molar" public_interface="in" name="L0"/>
    <variable units="number_per_cell" public_interface="in" name="R0"/>
    <variable units="first_order_rate_constant" public_interface="in" name="kr"/>
    <variable units="second_order_rate_constant" public_interface="in" name="kf"/>
    <variable units="first_order_rate_constant" public_interface="in" name="ku"/>
    <variable units="second_order_rate_constant" public_interface="in" name="kc"/>
    <variable units="number_per_cell" public_interface="in" name="r"/>
    <variable units="minute" public_interface="in" name="time"/>
    <math xmlns="http://www.w3.org/1998/Math/MathML" id="1">
      <apply> ... </apply>
    </math>
  </component>

  <component name="c2" cmeta:id="c2">
    <variable units="molar" public_interface="out" name="c2" initial_value="0.0"/>
    <variable units="molar" public_interface="in" name="c"/>
    <variable units="molar" public_interface="in" name="c_star"/>
    <variable units="number_per_cell" public_interface="in" name="R0"/>
    <variable units="first_order_rate_constant" public_interface="in" name="kr"/>
    <variable units="first_order_rate_constant" public_interface="in" name="kc_plus"/>
    <variable units="first_order_rate_constant" public_interface="in" name="kc_minus"/>
    <variable units="second_order_rate_constant" public_interface="in" name="kf"/>
    <variable units="first_order_rate_constant" public_interface="in" name="ku"/>
    <variable units="second_order_rate_constant" public_interface="in" name="kc"/>
    <variable units="minute" public_interface="in" name="time"/>
    <math xmlns="http://www.w3.org/1998/Math/MathML" id="2"> ... </math>
```

FIGURE 14.5

CellML pathway encoding.

14.5.1 ORA

Over Representation Analysis (ORA) [358]: ORA methods start from the preliminary candidate list found among user-selected lists of biological components through traditional differential expression analyses. Subsequently, the enrichment is computed using an iterative methodology evaluating the overlaps between the candidate genes and the genes belonging to the current pathway. Finally, statistical tests such as *chi-square* or hypergeometric or binomial distributions are applied to evaluate the reactome.all_species.interactions.psi-mitab.txt

```
#ID(s) interactor A	ID(s) interactor B	Alt. ID(s) interactor A	Alt. ID(s) interactor B	Alias(es) interactor A	Alias(es) interactor B
Interaction detection method(s)	Publication 1st author(s)	Publication Identifier(s)	Taxid interactor A	Taxid interactor B	Interaction type(s)
Source database(s)	Interaction identifier(s)	Confidence value(s)	Expansion method(s)	Biological role(s) interactor A	Biological role(s)
interactor B	Experimental role(s) interactor A	Experimental role(s) interactor B	Type(s) interactor A	Type(s) interactor B	Xref(s) interactor A
Xref(s) interactor B	Interaction Xref(s)	Annotation(s) interactor A	Annotation(s) interactor B	Interaction annotation(s)	Host
organism(s)	Interaction parameter(s)	Creation date	Update date	Checksum(s) interactor A	Checksum(s) interactor B	Interaction Checksum(s)
Negative	Feature(s) interactor A	Feature(s) interactor B	Stoichiometry(s) interactor A	Stoichiometry(s) interactor B	Identification method
participant A	Identification method participant B
uniprotkb:Q9Y287 uniprotkb:Q9Y287 reactome:R-HSA-976878|ENSEMBL:ENSG00000136156|ENSEMBL:ENSP00000497221|ENSEMBL:ENST00000647800|ENSEMBL:ENSP00000367811|
refseq:NP_068839.1|ENSEMBL:ENST00000378549 reactome:R-HSA-976878|ENSEMBL:ENSG00000136156|ENSEMBL:ENSP00000497221|ENSEMBL:ENST00000647800|
ENSEMBL:ENSP00000367811|refseq:NP_068839.1|ENSEMBL:ENST00000378549  reactome:"ITM2B(244-266)"(name)|reactome:ABri/ADan amyloid peptide(name)|
reactome:Integral membrane protein 2B(name)|reactome:ITM2B_HUMAN(name)|reactome:ITM2B(name)  reactome:"ITM2B(244-266)"(name)|reactome:ABri/ADan amyloid
peptide(name)|reactome:Integral membrane protein 2B(name)|reactome:ITM2B_HUMAN(name)|reactome:ITM2B(name)	psi-mi:"MI:0364"(inferred by curator)
Gibson et al. (2004)|Vidal et al. (1999)  pubmed:14690516|pubmed:10391242  taxid:9606(Homo sapiens) taxid:9606(Homo sapiens) psi-mi:"MI:0915"(physical
association)	psi-mi:"MI:0467"(reactome)	reactome:R-HSA-976871	reactome-score:0.5	-	psi-mi:"MI:0499"(unspecified role)	psi-
mi:"MI:0499"(unspecified role)	psi-mi:"MI:0499"(unspecified role)	psi-mi:"MI:0499"(unspecified role)	psi-mi:"MI:0326"(protein) psi-
mi:"MI:0326"(protein)	go:"GO:0005576"(extracellular region)	go:"GO:0005576"(extracellular region)	go:"GO:0005576"(extracellular
region) -	-	pathway:R-HSA-977225	taxid:9606(Homo sapiens) -	2010/10/14	2019/11/16	-	-	-	false
sufficient to bind:244-266	sufficient to bind:244-266	0	0	psi-mi:"MI:0364"(inferred by curator)	psi-mi:"MI:0364"(inferred by
curator)
uniprotkb:P37840 uniprotkb:P37840 reactome:R-HSA-976954|ENSEMBL:ENSP00000421485|ENSEMBL:ENSP00000378437|ENSEMBL:ENST00000505199|ENSEMBL:ENSP00000484044|
ENSEMBL:ENSP00000378442|ENSEMBL:ENSP00000396241|ENSEMBL:ENST00000673718|ENSEMBL:ENST00000618500|ENSEMBL:ENST00000345009|ENSEMBL:ENSP00000500990|
ENSEMBL:ENST00000336904|ENSEMBL:ENST00000394989|ENSEMBL:ENST00000394986|ENSEMBL:ENSP00000426955|refseq:NP_000336.1|ENSEMBL:ENSP00000338345|
ENSEMBL:ENST00000420646|ENSEMBL:ENSP00000422238|ENSEMBL:ENSP00000378440|ENSEMBL:ENSG00000145335|ENSEMBL:ENST00000508895|ENSEMBL:ENST00000506244|
refseq:XP_016864052.1|refseq:XP_016864051.1|refseq:NP_001139527.1|refseq:NP_001139526.1|refseq:XP_011530510.1|refseq:NP_009292.1|ENSEMBL:ENSP00000343683|
ENSEMBL:ENST00000394991	reactome:R-HSA-976954|ENSEMBL:ENSP00000421485|ENSEMBL:ENSP00000378437|ENSEMBL:ENST00000505199|ENSEMBL:ENSP00000484044|
ENSEMBL:ENSP00000378442|ENSEMBL:ENSP00000396241|ENSEMBL:ENST00000673718|ENSEMBL:ENST00000618500|ENSEMBL:ENST00000345009|ENSEMBL:ENSP00000500990|
ENSEMBL:ENST00000336904|ENSEMBL:ENST00000394989|ENSEMBL:ENST00000394986|ENSEMBL:ENSP00000426955|refseq:NP_000336.1|ENSEMBL:ENSP00000338345|
ENSEMBL:ENST00000420646|ENSEMBL:ENSP00000422238|ENSEMBL:ENSP00000378440|ENSEMBL:ENSG00000145335|ENSEMBL:ENST00000508895|ENSEMBL:ENST00000506244|
refseq:XP_016864052.1|refseq:XP_016864051.1|refseq:NP_001139527.1|refseq:NP_001139526.1|refseq:XP_011530510.1|refseq:NP_009292.1|ENSEMBL:ENSP00000343683|
```

FIGURE 14.6

PSI-MITAB pathway encoding.

over-representation. As a result, ORA methods provide a list of relevant pathways ordered according to their computed enrichment scores.

14.5.2 GSEA

Gene Set Enrichment Analysis (GSEA) [359]: GSEA approaches rely on a three-step methodology to calculate the enrichment using the entire gene list, including expression values, fold changes, p-values, etc. The first step involves figuring out the statistically most relevant genes. Gene significance is computed through ANOVA, fold change, t-statistic, log-likelihood ratio, t-test, and the Z-score. The second step consists of aggregating each genes' significance into pathway statistics. Kolmogorov–Smirnov, Wilcoxon sum rank, and chi-squared test statistical methods are employed to compute the pathway relevance. The final step regards the assessment of the pathway's statistical relevance. Statistical pathway relevance is assessed, permuting only genes' class labels (e.g., phenotype) into the pathway or permuting all the genes' class labels for each pathway and comparing the genes in the pathway with the genes of interest. GSEA, as a result, produces a summary enrichment score for a gene set.

14.5.3 TEA

Topological Enrichment Analysis (TEA) [360]: TEA methods are essentially the same as the ORA and GSEA methods. The main difference is that TEA combines the fold-change and p-value gene values with the pathway's topological information. Pathways are described as a graph, where nodes represent the pathway's components, e.g., genes, proteins, and gene products, and the edges provide information about the interactions among constituents, the type of interaction, e.g., activation or inhibition, along with topological information where they interact, e.g., nucleus, membrane, etc. This data is available in several pathway databases. The goal of the TEA approaches

```xml
<?xml version='1.0' encoding='utf-8' standalone='no'?>
<sbml xmlns="http://www.sbml.org/sbml/level3/version1/core" level="3" version="1">
  <notes>
    <annotation>
      <p xmlns="http://www.w3.org/1999/xhtml">SBML generated from Reactome version 77 on
      7/8/21 7:51 PM using JSBML version 1.5.</p>
    </annotation>
  </notes>
  <model id="pathway_10107048" metaid="metaid_0" name="Metabolism of nucleotides">
    <notes>
      <p xmlns="http://www.w3.org/1999/xhtml"> ... </p>
    </notes>
    <annotation>
      <rdf:RDF xmlns:rdf="http://www.w3.org/1999/02/22-rdf-syntax-ns#"
      xmlns:dc="http://purl.org/dc/elements/1.1/"
      xmlns:vCard="http://www.w3.org/2001/vcard-rdf/3.0#"
      xmlns:dcterms="http://purl.org/dc/terms/"
      xmlns:bqbiol="http://biomodels.net/biology-qualifiers/">
        <rdf:Description rdf:about="#metaid_0">
          <dc:creator>
            <rdf:Bag>
              <rdf:li rdf:parseType="Resource">
                <vCard:N rdf:parseType="Resource">
                  <vCard:Family>Shorser</vCard:Family>
                  <vCard:Given>Solomon</vCard:Given>
                </vCard:N>
                <vCard:ORG rdf:parseType="Resource">
                  <vCard:Orgname>OICR</vCard:Orgname>
                </vCard:ORG>
              </rdf:li>
            </rdf:Bag>
          </dc:creator>
          <dcterms:created rdf:parseType="Resource">
            <dcterms:W3CDTF>2021-05-21T20:34:56Z</dcterms:W3CDTF>
          </dcterms:created>
          <bqbiol:is>
            <rdf:Bag>
              <rdf:li rdf:resource="https://identifiers.org/reactome:R-BTA-15869" />
              <rdf:li rdf:resource="https://identifiers.org/GO:0055086" />
            </rdf:Bag>
          </bqbiol:is>
        </rdf:Description>
      </rdf:RDF>
    </annotation>
    <listOfCompartments>
      <compartment constant="true" id="compartment_7660" metaid="metaid_2"
      name="nucleoplasm" sboTerm="SBO:0000290">
        <annotation>
          <rdf:RDF xmlns:rdf="http://www.w3.org/1999/02/22-rdf-syntax-ns#"
          xmlns:bqbiol="http://biomodels.net/biology-qualifiers/">
            <rdf:Description rdf:about="#metaid_2">
              <bqbiol:is>
                <rdf:Bag>
                  <rdf:li rdf:resource="https://identifiers.org/GO:0005654" />
```

FIGURE 14.7

SBML pathway encoding.

to explicitly reveal the interaction among the most relevant genes of interest and the pathway's genes, highlighted by the pathway topology. Users must follow some best-practice rules when defining their pathway enrichment analysis to perform pathway enrichment by using tools based on:

- ORA methods—the users must provide a ranked or not gene list, a suitable threshold value, and the proper pathway database as inputs.

- pathway database as inputs. Setting the threshold value is unnecessary because GSEA methods score genes before computing enrichment.
- PEA methods—the users must provide a gene list, pathway database, and pathway topology as inputs.

14.6 Pathway enrichment analysis tools

Pathway enrichment software tools are available as stand-alone applications, web-based applications, and program libraries. The first two categories are usually more straightforward to use since they do not require analytical skills or programming capabilities. Conversely, the main benefits of using pathway enrichment software tools developed as programming libraries are the potential personalization of every step of the analysis and the feasibility to automate all the process steps through scripting analysis pipelines. Deciding between software platforms and program libraries may be related to the users' skills and the amount of data needing analysis.

The following sections present a description of the characteristics of the principal pathway enrichment software tools to provide guidelines to simplify the choice of PEA methods to analyze the available DEGs data.

14.6.1 ORA software tools

The following lists some well-known ORA software tools.

14.6.1.1 GO-Elite

GO-Elite [358] supports researchers to perform ORA using structured ontology annotations, pathway databases, or biological set of biological entities to identify pathways or ontology terms to describe a group of genes of interest. GO-Elite can be downloaded freely as a stand-alone application directly from http://www.genmapp.org/go_elite. It is also available as Web Service[2] and as Cytoscape plugin, making it compatible with macOS, Windows, and Linux Operating Systems. GO-Elite needs two input files to perform enrichment analysis. One input file contains the list of the genes of interest, and the second one is the expression data file in *GenMAPP-CS* format. GO-elite allows users to export results in multiple file formats, such as over-representation analysis results, pruned ORA results, gene annotation file, gene ranking file, input gene pathway associations file, and comparison of ontology and pathway terms. GO-Elite enrichment scores are ranked according to a *Z-score*. Enrichment results are estimated using a normalized version of the *hypergeometric distribution* and *Fisher's Exact Test*. To perform pathway enrichment analysis, GO-Elite can retrieve pathways data from the following databases: WikiPathways [354], KEGG [361], and PathwayCommons [351].

[2] http://webservices.cgl.ucsf.edu/opal2/CreateSubmissionForm.do?serviceURL=http://localhost:8080/opal2/services%2FGOEliteService.

14.6.1.2 BiP

BioPAX-Parser (BiP) [147] assists researchers in identifying appropriate pathways to delineate the role of a set of interest genes. BiP performs pathway enrichment analysis by just uploading a list of proteins or genes of interest, along with the pathways to investigate encoded in BioPAX [355] format. BioPAX is a metalanguage defined in OWL (Web Ontology Language) and coded in the RDF/XML (Resource Description Framework / eXtensible Meta Language) format. BiP is freely available for download as a stand-alone application at https://gitlab.com/giuseppeagapito/bip. BiP is fully developed in Java, making it platform-independent, which means it can be executed on all operating systems compatible with Java. BiP requires as input a plain text file containing the list of genes or proteins of interest to be investigated. BiP enrichment results can be exported in tabular format. Pathway enrichment analysis is achieved by implementing a customized version of the *hypergeometric* function, along with multiple statistical correctors such as False Discovery Rate (*FDR*) and *Bonferroni*. BiP can perform pathway enrichment using pathway information from all the available pathway databases compliant with the BioPAX format.

14.6.1.3 PathDIP

PathDIP [362] comprises core pathways from major curated pathway databases and pathways predicted based on orthology and by using physical protein interactions. PathDIP helps researchers in performing ORA on structured ontology annotations, pathway databases, or sets of biological entities to identify pathways or ontology terms to describe a set of genes of interest. PathDIP is publicly available at http://ophid.utoronto.ca/pathDIP. It is a web application making it platform-independent and compatible with macOS, Windows, Linux, and all the other available OS. In addition, pathDIP is available as an Application Program Interface (API) in Java, R, or Python. PathDIP needs an input list of proteins or genes and the selected annotation sets to discover significantly enriched pathways. To calculate enrichment scores, PathDIP applies *Fisher's Exact Test*, followed by correction for multiple hypothesis testing by two different methods, *Bonferroni* and *FDR*. Enrichment results and detailed annotations for the input list can be exported in tabular format. pathDIP integrates ASCN2 [363], BioCarta [364], EHMN [365], HumanCyc [366], INOH [367], IPAVS [368], KEGG [361], NetPath [369], OntoCancro [370], Panther [371], PharmGKB [372], PID [373], RB-pathways [374], Reactome [375], Signalink2.0 [376], SIGNOR2.0 [377], SMPDB [378], SPIKE [379], STKE [380], System-biology.org [381], UniProt Pathways (https://www.uniprot.org/help/pathway), and WikiPathways [354] to perform pathway enrichment.

14.6.2 GSEA software tools

The following lists some well-known GSEA software tools.

14.6.2.1 g:Profiler

g:Profiler [382] helps researchers to gain insight into a genes list using the Gene Set Enrichment Analysis (GSEA). g:Profiler maps genes to known functional information sources and detects statistically significantly enriched pathways, more than would be expected by chance. g:Profiler is available freely as a web application at https://biit.cs.ut.ee/gprofiler/ and implemented in *Python 3.6*. In addition, g:Profiler is available as both Python and R client libraries and as an API. The default input data format of g:Profiler is a list of genes or proteins. The input gene list can be either unordered or ordered (as the default options list is considered ordered). For instance, a gene list can include mixed types of gene ID (proteins, transcripts, and microarray ID), SNPs, or chromosomal intervals. g:Profiler provides the computation results in three separate tabs, i.e., *Results*, *Detailed Results*, and *Query Info*. g:Profiler can show the users' enrichment analysis results through an interactive Manhattan plot or an interactive result table containing the enriched terms. Results can be exported as images or Comma Separated Values (CSV) files. The enrichment of the input gene list is assessed using the well-known *hypergeometric* test to correct results obtained employing multiple testing, e.g., hypergeometric test, g:Profiler uses the *False Discovery Rate* (FDR) and *Bonferroni* correctors. To perform pathway enrichment, g:Profiler retrieves information coming from KEGG [361], WikiPathways [354], and Reactome [375] pathway databases.

14.6.2.2 GSEA

GSEA [359] helps researchers to perform pathway enrichment using gene expression data sets. It is particularly suitable when expression data are available for all the genes under investigation. GSEA is freely available as a stand-alone application. To download the GSEA software, users have to register at https://www.gsea-msigdb.org/gsea/login.jsp. In addition, GSEA is available as a Java and R application. In this way, GSEA is compatible with all the operating systems supporting Java and R programming languages. To start pathway enrichment analysis, GSEA needs four types of data files: an *expression dataset* in any of the following formats (RES, GCT (Gene Cluster Text), PCL, or txt) file; a *phenotype labels* file in CLS format; a *gene sets* defined using the (GMX (Gene MatriX) or GMT (Gene Matrix Transposed)) file format; and a (*CHIP: Chip file*) file format with microarray annotations. All four file types have to be tab-delimited. The functional enrichment of the input gene list is evaluated using the enrichment score. To reduce the errors due to multiple testing, GSEA uses the *FDR* and *Bonferroni* correctors. As a result, GSEA produces various types of reports: enrichment in phenotypes, detailed enrichment results, and gene set details report, to cite just a few. All the results can be exported as hypertext-markup language (HTML) or xlsx (Excel) files. GSEA can retrieve information from KEGG [361], BioCarta [364], PID [373], and Reactome [375] pathway databases. Moreover, users can employ any other pathways database compliant with the GMT format.

14.6.3 TEA software tools

The following lists some well-known TEA software tools.

14.6.3.1 PathNet

Pathway analysis using Network information (PathNet) [383] exploits topological information present within the pathways and differentially expressed genes to identify relevant enriched pathways associated in the context of gene expression data. PathNet is a collection of R functions for pathway enrichment analysis employing topological information embedded in the pathways. To perform topological enrichment analysis, PathNet combines all pathways under consideration into a pooled pathway. The interactions among genes in the pooled pathway are represented as a binary adjacency matrix, where 1 represents the presence of the interaction and 0 the absence. Given the network, PathNet computes the node relevance score using *Fisher's* exact test. PathNet requires the following types of input data: differential expression levels; adjacency matrix; and pathway information to perform topological enrichment analysis. Enrichment results can be downloaded as a couple of plain-text files. PathNet can estimate topology information from KEGG [361] pathway database only.

14.6.3.2 TPEA

Topology-based pathway enrichment analysis (TPEA) [384] computes the significance of nodes based on their upstream/downstream positions and the degrees within the pathways. TPEA is available as an R application at the following address: https://cran.r-project.org/web/packages/TPEA/. TPEA is compatible with all the OS supporting R programming languages. Input data sets should contain differentially expressed genes or any other type of gene data to be suitable for topological enrichment. The input gene data sets must be encoded using the *"Entrez ID"* format. If not, it is mandatory to translate the input genes into the *Entrez ID* format. TPEA computes the area under the enrichment curve (AUEC). AUEC evaluates the cumulative weighted node score obtained from the input genes to assess the pathways enrichment. Results can be saved in a tabular format. TPEA uses only the KEGG database to compute the pathway enrichment analysis.

CHAPTER

Knowledge extraction from biomedical texts

15

CONTENTS

Abstract

The daily production of textual content suitable for memorization and analysis is growing rapidly in the biomedical sector. The incredible productivity rate of textual content represents a valuable resource for extracting knowledge potentially useful for decision-making purposes. Text mining refers to the process of extracting knowledge from text. In this chapter a brief overview of the text mining process is presented, and some among the most popular text mining tasks in the biomedical context are discussed.

Keywords

Text mining, NLP, Named entity recognition, Topic modeling, Sentiment analysis

15.1 Introduction

The daily production of textual content suitable for memorization and analysis is growing rapidly in the biomedical sector. Some examples include, but are not lim-

Artificial Intelligence in Bioinformatics. https://doi.org/10.1016/B978-0-12-822952-1.00025-5

ited to: Electronic Health Records (EHR), the scientific literature, narrative medicine, open-ended questionnaires, etc. Suffice it to say that, to date, almost two years after the start of the COVID-19 pandemic, PubMed reports that there are approximately 222,000 scientific articles that reference SARS-CoV-2 and COVID-19.

The incredible productivity rate of textual content represents a valuable resource for extracting knowledge potentially useful for decision-making purposes. Hence, the need to design dynamic and scalable algorithms specifically designed for the discovery of new knowledge from textual sources represents one of the greatest research challenges. Introduced by Feldman et al. [385], Knowledge Discovery in Text (KDT) or text mining refers to the process of extracting high-level information from text.

From a data science perspective, text mining customizes the data science principles and techniques in a way suitable for dealing with text, e.g., the general pipeline of a text mining application, or the four identified "super-problems"—classification, regression, association rules, and clustering—in which basically every text mining problem can be conceptually categorized.

However, the intrinsic characteristics of textual data strongly distinguish them from other data structures, e.g., relational data. In fact, in a text mining system the input data consists of collections of unstructured documents that can be analyzed at various levels of representations: the word-level, sentence-level, and document-level. Moreover, representing each single word as a distinct feature leads to a highly dimensional and sparse representation of the problem.

Therefore, specific techniques born at the intersection of various fields such as data mining, machine learning, statistical methods, information retrieval, and natural language processing (NLP) have been developed for extracting knowledge from text.

The chapter introduces some background on text mining and then presents some of the most popular text mining tasks in the biomedical domain.

15.2 A primer on text analysis

The following section introduces and discusses a typical analysis pipeline for text mining [386].

15.2.1 Data collection

The first step of a text mining process is the collection of textual data. In many text-mining real-world applications, the textual data of interest may already be provided or may be strictly related to the problem description, e.g., in EHR. Several textual data collections, also called corpora, are publicly available for the research and the development of general purpose text mining algorithm, as well as specific domain oriented corpus.

15.2.2 Text pre-processing

A key point in text mining consists of the encoding of textual input into a meaningful representation that should be suitable for the application of standard data mining procedures and should provide structured and numerical features. Before introducing multiple approaches to text representation, a first step of data pre-processing is needed.

Common pre-processing steps in text mining are treated in the next sections:

15.2.3 Text cleaning

This removes anchors or tags, filters noise (e.g., non-ASCII characters, etc) which is especially needed when text is extracted from web. If input data comes from social networks, pre-processing requires other several steps, such as online text cleaning (like removing URLs, HTML tags, or the retweet tag), expanding abbreviations or acronyms, handling or removing emoticons, and replacing or removing repeated characters.

15.2.4 Tokenization

Tokens are basic text units with a semantic meaning and, in general, are assumed to be words. Tokenization, i.e., the automatic process of extracting tokens from a given textual input, presents some challenges related to the detection of word boundaries. Even though it may be considered a trivial task for humans, tokenization is a tricky problem. Various strategies for dealing with apostrophes and hyphens, capitalization, punctuation, numbers, and alphanumeric strings to set a maximum length for characters composing tokens that deal with non-alphanumeric characters lead to a slightly different text segmentation output [385]. Another difference is that some tokenization algorithms filter punctuation, while others consider a punctuation mark as a separate token. In general, an important remark is that the output of a tokenization is a list of detected tokens.

15.2.5 Filtering

The most commonly used words in a language are the ones that usually provide the least amount of information. Hence, a common practice is to filter this set of less informative words, known as stop words. There is not a universal list of stop words because, in general, they are language and task dependent.

15.2.6 Lemmatization and stemming

Lemmatization refers to the process of identifying words having common morphological root and replacing them with the same representative token. In general, lemmatization approaches are based on handcrafted set of rules that takes into account the specific part-of-speech of a word, or are lexicon-based. A simpler form of lemmatization is stemming, in which suffixes are stripped from the end of a word by using a set of rule for suffix stripping. While the latter approach is very easy to implement, in general it is too aggressive and may change the semantic meaning of a word. For example, "hoping" may be replaced with "hop" instead of "hope" (over-stemming) and, on the contrary, irregular verbs such as for "be" or "run" and "ran" may not be recognized.

15.2.7 Text representation

The representation of the text in a numerical form suitable for computation is one of the fundamental problems in information retrieval that occupies a central importance also in the text mining field.

The most used textual representation models refer to the Vector Space Model (VSM), in which each document is represented as a vector in an N-dimensional space and a suitable distance between vectors is defined that, in general, reflects the similarity of documents.

15.2.7.1 Text vectorization

Three are the main elements to build a vector space model: i) the identification of a coordinate system in the vector space; ii) the determination of a weighing scheme that, given a document, determines its components with respect to the chosen coordinate system; iii) the identification of an appropriate vector distance that maintains semantic information.

The **Bag-of-Words Model** (BOW) considers as a feature the set of distinct words so feature space has as many dimensions as the number of different words [387].

The most commonly used term-weighing models in BOW representation are:

- Boolean models, in which $w_i = 1$ if the term t_i is present at least once in the document d_l, otherwise $w_i = 0$;
- Term frequency-inverse document frequency (tf-idf). When considering a collection of D documents, let $f_d(t)$ indicate the frequency of a term t in the document d and $f_{\mathbb{D}}(t)$ the frequency of the term t in the document collection. Under the assumption that the frequency of a term is the raw count of the presence of the term in the document or in the document collection, the tf-idf is computed as follows:

$$tf - idf(t, d, \mathbb{D}) = f_d(t) * log \frac{|D|}{f_{\mathbb{D}}}(t).$$

Tf-idf is preferred to basic term frequency because, in the latter, common terms, which are not as informative as less common terms, will be over-weighted. On the contrary, *tf-idf* is proportional to the number of occurrences of a fixed word in the document, and it is inversely proportional to the frequency of that word in the document collection. Therefore heavy weights are assigned to terms that are frequently mentioned in a specific document but not in the whole collection.

In general, BOW model is very popular because of its simplicity, but one of its main disadvantages is the loss some syntactic information of the text, e.g., ordering.

The sentence "The service was good but the food was awful" shares the same BOW representation of "The food was good but the service was awful" even if it's clear that they have different syntax and very different meanings.

Therefore, several text representation models have been proposed in the literature to take into account multiple features such as **N-grams**. An N-gram is a sequence of N contiguous terms. In general N-grams representations may follow a Boolean model. The major advantage of an N-grams representation is to capture relationships among the current word and the previous N words. However, by modeling the sequence of occurring word as a Markov chain, the major drawback is that N-grams lead to a sparser and higher-dimensional representation compared to the BOW model.

15.2.7.2 Distributed representation of text

The representations presented so far are highly interpretable because they are massively based on hand-coded rules; however, they usually lead to a sparse and highly dimensional representation that may be redundant and lead to losing syntactic information during computation. For instance, if we consider a corpus with 50,000 terms after text normalization, it will lead to a 50,000 vector-space dimensional representation, at the least. An alternative representation that has become a state-of-the art approach in recent years is to exploit neural network architectures to learn a distributed representation of words. A distributed representation is a features representation that makes use of low-dimensional and dense vectors to implicitly represent the syntactic or semantic characteristics of the language.

The distributed representation approach is also known as **word embedding**, the first and most popular of which is word2vec. Although distributed text representations are generally learned when simulation a specific NLP task, the increase in popularity of various algorithms has discovered that models pre-trained on very large corpus can learn universal linguistic representations.

Over the years, the typology of representations has evolved. Initially, the goal was to learn the embedding of words without context. Models currently aim to learn word embedding that captures higher-level concepts in context [388] through the use of architectures known as transformers [389], the most popular of which is certainly BERT [390].

Several pre-trained language models have been proposed in the biomedical field, such as BioBert [391], ClinicalBert [392], MedBert [393], and PubMedBert [394]. The standard approach is to obtain biomedical pre-trained language models by pre-training general BERT model on biomedical texts [395].

15.3 Biomedical text mining tasks

In this section the most popular text mining tasks and their application in the biomedical field are discussed.

15.3.1 Named entity recognition

In the biomedical field, one of the most relevant TM task concerns the recognition of entities from raw text. Generally, a named entity can be defined as the word or set of words that identify an entity. For example, the set of words "United States" represents an entity of the type state. The task that aims to recognize named entities in multiple pieces of textual data is called Named Entity Recognition (NER).

The extraction of named entities in the biomedical field has various applications, which include the recognition of biological entities or concepts in the scientific literature, e.g., the names of genes and proteins, the recognition of entities or medical concepts (e.g., the name of a disease, anatomical region) in clinical reports, the recognition of sensitive data in clinical reports for purposes of anonymization or temporal expressions, etc. Furthermore, the NER task is essential for another very interesting task in the biomedical field, namely, the extraction of relationships, e.g., semantic relationships between biological entities, such as protein interactions or extraction of relationships between medical concepts (e.g., disease symptom) and event extraction [396].

15.3.2 Text summarization

The purpose of the automatic text summarization task is to produce a summary of a text (single-document) or of a collection of texts (multi-document) that preserves the general meaning of the document or documents by enhancing the content of the most important information. One of the major applications that text summarization finds in the biomedical field is related to the summarization of the biomedical literature, providing domain experts with tools for determining clear information in the large amounts of information embedded in textual resources. The classical division of text summarization approaches distinguishes the techniques in which the text summary can be constructed by considering the most important sentences of the source text, i.e., extractive text summarization, from techniques more similar to human synthesis, in which the text that summarizes the source document is generated from scratch, interpreting and capturing only the essential content in the text. The majority of text summarization approaches are extractive.

15.3.3 Topic modeling

Similar in scope to text summarization but different in approach, another classic text mining task is topic modeling A topic can be seen as a latent variable in the text. It is composed by a collection of representative words in a document that helps identify the subject or the subjects of a document. Topic modeling is the task of identifying topics in collections of documents. The basic assumption of thematic models is that each document consists of a mixture of topics.

Latent Dirichlet Allocation (LDA) [397] is a generative probabilistic model commonly used for the identification of latent topics in textual corpora. The assumption under an LDA model is that each document in a corpus can be modeled as a mixture of a finite number of topics with a certain probability, while a distribution over words can characterize a topic.

15.3.4 Sentiment analysis

Sentiment Analysis (SA) refers to that computational area devoted to the automatic extraction of opinion and emotions, mainly, but not only, from textual sources.

SA methods, techniques, and tools have been developed in parallel in various research areas, resulting in the increase in the number of different terminologies that reflect some subtle differences that strongly depend on the application domain, e.g., SA, opinion mining, opinion extraction, sentiment mining, emotion analysis, and review mining, among others. In recent years, with the aim of considering all these research areas "as representing a unified body of work", SAs has been considered as an umbrella grouping many natural language processing problems. Therefore, regardless of the application context, the main goal of a SA task is to predict the sentiment orientation or the sentiment intensity of a given text according to its content.

The most common task is **polarity detection** that aim is to determine whether a text contains a positive, negative, or in some cases neutral opinion, i.e., its polarity [309,398].

The usefulness of sentiment analysis applications in the medical field is demonstrated by various applications that raise the desire to formalize a prediction algorithm of adverse/negative events.

In particular, SA has been applied to improve the early detection of a whole range of depression-related disorders, such as post traumatic stress, bipolar disorder, and seasonal affective disorder. Examples of such work can be found respectively in [399–403].

CHAPTER

Artificial intelligence in bioinformatics: issues and challenges

16

CONTENTS

Abstract

This chapter recalls the evolution of bioinformatics through several definitions of the term “bioinformatics” through the years, from its first definition in 1970 until now. Then, the chapter summarizes the main challenges of artificial intelligence that are particularly important when applying artificial intelligence in bioinformatics and that may prevent its use in real applications.

Keywords

Artificial intelligence bioinformatics, Explainability in artificial intelligence, Bias in artificial intelligence, High performance computing, Privacy and security of AI, Privacy and security regulations, Medical informatics, Biomedical informatics

Artificial Intelligence in Bioinformatics. https://doi.org/10.1016/B978-0-12-822952-1.00026-7

16.1 Introduction

Bioinformatics is an interdisciplinary science that involves biology, informatics, computer science, mathematics, statistics, physics, and other disciplines. It is an evolving discipline, and its evolution was initially driven by the emergence of novel high-throughput technological platforms able to generate huge volumes of molecular data. A second driver of the evolution was its broader application to more disciplines, such as medicine and health sciences, other than biology. Finally, a third evolutionary aspect is related to the application of machine learning and artificial intelligence methods to the immense knowledge base constituted by biological data banks, biological experimental data, and, finally, human clinical data.

Moreover, the fields where bioinformatics is applied are connected to additional terms that involve informatics, such as medical informatics and biomedical informatics. A clear distinction between *bioinformatics*, *medical informatics*, and *biomedical informatics* is reported on the EBI website[1] and can be summarized as follows:

"Bioinformatics is different from medical informatics, that regards the design, development, and application of information and communication technologies in healthcare services. On the other hand, biomedical informatics is another discipline that is between bioinformatics and medical informatics. It regards the effective uses of biomedical data for scientific research, problem solving and decision making, with the main aim to improve human health."

The most recent, and probably final, evolution of bioinformatics is certainly due to an increasing use of artificial intelligence and its deep application in medicine. As explained in the book, the application of artificial intelligence in bioinformatics (that we can term as *artificial intelligence bioinformatics*) has many positive aspects, but poses a series of challenges that are related not only to classical challenges and problems of artificial intelligence, in itself, but that are also related to the application fields were artificial intelligence bioinformatics is applied, such as medicine. The rest of the chapter recalls the evolutionary steps of bioinformatics and summarizes the main challenges of using artificial intelligence in bioinformatics

16.2 Evolution of bioinformatics

The evolution of bioinformatics can be traced by recalling the different definitions that were given to bioinformatics over time, from its first definition in 1970 until now, and that are reported in this section.

16.2.1 Origin of bioinformatics

The term bioinformatics was first coined by Paulien Hogeweg and Ben Hesper with the aim to describe *"the study of informatic processes in biotic systems."* The origi-

[1] https://www.ebi.ac.uk/training/online/courses/bioinformatics-terrified/what-bioinformatics/.

nal paper reporting the term was published in Dutch in 1970 [404]; a translation in English is now available [405]. The term started to be used when the first biological sequence was discovered and shared. Initially, bioinformatics methods thus focused on the comparison of linear sequences or 3-D structures.[2]

16.2.2 NIH-NHGRI definition of bioinformatics

The National Institute of Health (NIH)—National Human Genome Research Institute (NHGRI) defines bioinformatics in the following way[3]: *"Bioinformatics is a subdiscipline of biology and computer science concerned with the acquisition, storage, analysis, and dissemination of biological data, most often DNA and amino acid sequences. Bioinformatics uses computer programs for a variety of applications, including determining gene and protein functions, establishing evolutionary relationships, and predicting the three-dimensional shapes of proteins."* From this definition, it is clear that the focus was on the acquisition, storage, analysis, and dissemination of biological data, mainly composed of DNA and amino acid sequences at that time.

16.2.3 SIB definition of bioinformatics

The Swiss Institute of Bioinformatics (SIB) reports this definition[4]: *"Bioinformatics is the application of computer technology to the understanding and effective use of biological and biomedical data. It is the discipline that stores, analyzes, and interprets the big data generated by life-science experiments, or collected in a clinical context. This multidisciplinary field is driven by experts from a variety of backgrounds: biologists, computer scientists, mathematicians, statisticians and physicists."* The SIB web site introduces this definition by stating that *"Nowadays, there is no shortage of data. But a different kind of problem has emerged. New technologies are producing data at an unprecedented speed. Indeed, so much data—and of such variety—that they can no longer be interpreted by the human mind alone"*, as continuously remembered in various points of this book. In this definition the focus is yet on the data and on the impossibility to deal with them with only the human mind, therefore, giving a central role to automatic processing through computer science.

16.2.4 EBI definition of bioinformatics

The European Bioinformatics Institute (EBI), reports the following definition[5]: *"Put simply, bioinformatics is the science of storing, retrieving and analysing large amounts of biological information. It is a highly interdisciplinary field involving many*

---

[2] https://www.historyofinformation.com/detail.php?id=5137.
[3] https://www.genome.gov/genetics-glossary/Bioinformatics.
[4] https://www.sib.swiss/what-is-bioinformatics.
[5] https://www.ebi.ac.uk/training/online/courses/bioinformatics-terrified/what-bioinformatics/.

different types of specialists, including biologists, molecular life scientists, computer scientists and mathematicians."

16.2.5 Other definitions of bioinformatics

Another definition was given in 2001 by Luscombe, Greenbaum, and Gerstein [406]. Their definition is quite following the older ones and is as follows. *"Bioinformatics is conceptualizing biology in terms of macromolecules (in the sense of physical-chemistry) and then applying "informatics" techniques (derived from disciplines such as applied maths, computer science, and statistics) to understand and organize the information associated with these molecules, on a large-scale."*

Another definition was given in 2002 by Bayat [407] that reports the following characteristics of bioinformatics:

- *"Bioinformatics is the application of tools of computation and analysis to the capture and interpretation of biological data.*
- *Bioinformatics is essential for management of data in modern biology and medicine.*
- *The bioinformatics toolbox includes computer software programs such as BLAST and Ensembl, which depend on the availability of the internet.*
- *Analysis of genome sequence data, particularly the analysis of the human genome project, is one of the main achievements of bioinformatics to date.*
- *Prospects in the field of bioinformatics include its future contribution to functional understanding of the human genome, leading to enhanced discovery of drug targets and individualised therapy."*

16.2.6 Artificial intelligence bioinformatics

We define **artificial intelligence bioinformatics** as the "deep use of Artificial Intelligence methods in several bioinformatics tasks, with special focus on the building of predictive models trained on Big Omics Data, eventually integrated with clinical data."

16.3 Challenges for artificial intelligence in bioinformatics

Many of the challenges of applying artificial intelligent methods in bioinformatics are directly related to well-known challenges of artificial intelligence itself. In a recent article in the magazine **upGrad**,[6] Pavan Vadapalli illustrates seven common challenges in artificial intelligence. In the following we summarize them with special focus on the application of artificial intelligence in bioinformatics.

[6] https://www.upgrad.com/blog/top-challenges-in-artificial-intelligence/.

16.3.1 High demand of computing power

A first challenge is the high power consumption of AI systems. Machine learning and deep learning methods that are at the basis of all AI systems require ever-increasing computational power. This requirement is faced by using computers with increasing power, which may be obtained by increasing the number of computer cores and by using, in a combined way, CPUs (Central Processing Units) and GPUs (Graphic Processing Units). Thus, running AI system requires supercomputers, which have higher costs, and computer scientists skilled in high-performance computing, both of which can be a challenge for small bioinformatics laboratories. Although exploiting cloud computing services may reduce hardware and systems costs, these are replaced by the cost of parallel processing developers.

16.3.2 Lack of trust and explainability

As explained in Chapter 6, one of the limiting factors for the adoption of AI systems is the unknown nature of how deep learning models predict the output, and how a certain decision can be explained in terms of features that are known to the domain expert. This is a very limiting factors in several domains and especially when bioinformatics results are used to take decisions in medicine.

16.3.3 Limited awareness of AI potentiality

The general public has a low awareness of the use or existence of artificial intelligence and how it is integrated into everyday applications. Moreover, often also computer scientists and decision makers have a limited awareness of the potential of AI, and how it could be used to improve their activities. A similar unawareness regards the availability of cloud services based on AI. This scarce knowledge and awareness about the potentiality of AI is also reflected in its slow adoption in bioinformatics.

16.3.4 Human vs AI precision

Many AI systems do not yet reach the precision of humans, preventing their diffusion into many domains where decision makers prefer to invest in human resources and not on AI development. Improving the precision of AI systems to the human level may require much fine tuning and especially big data sets for the training phase that require both computational power and skilled computer scientists. A possible solution to this problem is using service providers that can take care of the training phase of the AI system by using internal pre-trained models, but this approach is yet lacking in bioinformatics.

16.3.5 Data privacy and security

The enabling factors for training AI systems are: the machine learning method, the computational power, and the data to train the system. Although we live in the information era where huge volumes of data are daily produced, this data often are produced by (and belong to) millions of users. This poses two problems: it would be unethical for an AI system to use the data owned by millions of users, and it would be possible that those data are used for bad purposes. Moreover, concentrating in a single place the data of million users (e.g., health and medical data) may constitute a high risk of data leakage involving sensitive personal data. This problem is central to bioinformatics studies that use omics data obtained by human donors or patients. The main approaches to address these issues are the adoption of privacy and security laws that recently have been imposed at the national or continent level. Relevant examples of such regulations are: the European General Data Protection Regulation (GDPR)[7] and the Health Insurance Portability and Accountability Act (HIPAA).[8] The latter is a federal law of United States (US) driving the creation of national standards to protect sensitive patient health information. Other local regulations are the California Consumer Privacy Act (CCPA),[9] the Canadian Personal Information Protection and Electronic Documents Act (PIPEDA),[10] the Brazilian law on privacy—Lei Geral de Proteção de Dados (LGPD), and the South Africa Protection of Personal Information Act (POPI Act), among others.

Applying these regulations may be costly and may prevent the use of AI systems in bioinformatics. To avoid the concentration of massive data in a central server that may lead to data leakage in the case of cyberattacks, some AI systems train the data on small devices close to places where the data originated, without moving the data to a central server, and only the trained model is maintained centrally. This approach can exploit the recent developments in edge computing (also known as fog computing) [408], a distributed computing model where computation is moved to the edge of the network and is enabled by the increasing power of smartphones and sensors, such as the Internet of Things (IoT) devices. The work [409] surveys some machine learning systems that have been deployed at the edge of computer networks, moving AI to the edge.

16.3.6 The bias in AI

The good or poor prediction produced by an AI system mainly depends by the quality of data used for the training ("better data yield to better models"). Thus a key factor to train good models is to use good data. Of course, collecting good data is costly, and it requires a lot of preprocessing and human work. Even if data are properly cleaned and preprocessed, they can be biased, i.e., they can reflect the positions of a limited number of people with respect to religion, ethnicity, gender, community, and other social biases. Thus, although an AI system can be technically correct, it may make decisions that take into account those biases. A different formulation of this problem

---

7 https://gdpr.eu/what-is-gdpr/.

8 https://www.cdc.gov/phlp/publications/topic/hipaa.html.

9 https://oag.ca.gov/privacy/ccpa.

10 https://www.priv.gc.ca/en/privacy-topics/privacy-laws-in-canada/the-personal-information-protection-and-electronic-documents-act-pipeda/.

in bioinformatics is related to the large use of electronically inferred information versus manually curated data or data obtained from wet lab experiments. For instance, in the ontology domain, annotations with IEA (Inferred from Electronic Annotation) Evidence Code comprise the majority. Thus, bioinformatics analysis may be biased by the kind of data that may have a lower reliability with respect to experimental or manually curated data.

16.3.7 Data scarcity

Due to the recent regulations that may limit the collection and analysis of data (especially of personal ones), data scarcity is becoming an issue. Many companies attempt to train models with the few data that regulations permit to be collected. But, collecting few data is one of the main causes of the data bias problem because few data may mean that only a few examples are represented. Thus, the data scarcity problem is often a cause of the data bias problem.

Another source of data scarcity is the difficulty of collecting enough clinical data, either due to privacy regulations or often due to the difficulty to join clinical data coming from international studies because of the various national regulations or intrinsic complexity of data management on a large scale. A possible solution to this problem is the use of innovative data integration and machine learning approaches, such as those used by the Consortium for Clinical Characterization of COVID-19 by EHR (4CE),[11] where clinical data about COVID-19 were collected by more than 350 hospitals worldwide [410,411]. To face local privacy regulations, only locally aggregated data are shared and moved to a central server, and global data analysis is performed centrally on aggregated data. On the other hand, the 4CE project defined a series of R scripts with common quality control criteria that are made available through the docker technology, enabling locally a finer analysis of the disaggregated clinical data, which is further integrated on a global scale.

Even with those challenges, the benefits of using artificial intelligence in bioinformatics are overall superior to the problems described so far, so it is foreseeable that artificial intelligence will be more and more used in bioinformatics and related disciplines.

[11] The Consortium for Clinical Characterization of COVID-19 by EHR (4CE) is an international consortium for electronic health record (EHR) data-driven studies of the COVID-19 pandemic. The consortium includes sites in America, Europe, and Asia and is driven by the Harward Medical School. More information is available here: https://covidclinical.net/.

APPENDIX

Python code examples

Throughout this book, we use Python to perform some omics data analysis, employing the most suitable third-party packages that offer machine learning, an intelligent agent framework, and statistical support to analyze omics data in a simple, straightforward way.

This chapter will present what can be considered the tip of the iceberg of machine learning, agent-based modeling, and statistical omics data analysis. However, readers have the tools and knowledge to process the omics data with and without supervision, the minimum skills to define an intelligent agent, and the most suitable statistical functions to use with a particular type of omics data. Readers can configure omics data experiments to verify their validity and draw non-trivial conclusions.

A.1 Classification of omics data

To better illustrate how to use SVM to classify genomics data, we present a straightforward Python script to classify the *Wisconsin breast cancer* dataset available into the `scikit-learn` module. The breast cancer dataset is a binary classification dataset. It encompasses two classes, "Malignant" and "Benign", respectively, with a total of 569 samples. Samples labeled as "Malignant" are 212, whereas samples labeled "Benign" are 357. Let's analyze the script A.1 to understand how to train our *SVM* model, and next, how to verify its prediction accuracy.

```
1  import matplotlib.pyplot as plt
2  import numpy as np
3  import pandas as pd
4  import seaborn as sns
5  from sklearn.model_selection import train_test_split
6  from sklearn.datasets import load_breast_cancer
7  from sklearn.svm import SVC
8  from sklearn.metrics import classification_report, confusion_matrix
9  from sklearn.preprocessing import MinMaxScaler, StandardScaler
10 cancer = load_breast_cancer()
11
12 #split the dataset using the 80% of the instances for training and
13 # the remainder 20% as testing "test setest_size=0.2, train_size=0.8"
```

```
14 X_train, X_test, y_train, y_test = train_test_split(cancer.data,
15 random_state=20, test_size=0.2, train_size=0.8)
16 print(X_train.shape)
17 print(X_test.shape)
18 #SVM model creation by using kernel function C=100
19 svm = SVC(C=100)
20 #SVM model training start
21 svm.fit(X_train, y_train)
22 print("Test set accuracy: {:.2f}".format(svm.score(X_test,y_test))) #
       SVM model accuracy evaluation
23 #Using the trained model to make a prediction using the test set
24 y_predict = svm.predict(X_test)
25 #model accuracy checking by using the confusion matrix
26 cm = np.array(confusion_matrix(y_test, y_predict, labels=[1, 0]))
27 confusion_mat = pd.DataFrame(cm, index=['sick', 'healthy'], columns=['
       predictedSick', 'predictedHealthy'])
28 print(confusion_mat)
29 sns.heatmap(confusion_mat, annot=True)
30 plt.show()
31 print(classification_report(y_test, y_predict))
32 #Normalize training data
33 scaler = MinMaxScaler()
34 scaler.fit(X_train)
35 X_train_scaled = scaler.transform(X_train)
36 X_test_scaled = scaler.transform(X_test)
37 print("Feature min values before scaling:\n {}".format(X_train.min(
       axis=0)))
38 print("Feature max values before scaling:\n {}".format(X_train.max(
       axis=0)))
39 print("Feature min values before scaling:\n {}".format(X_train_scaled.
       min(axis=0)))
40 print("Feature max values before scaling:\n {}".format(X_train_scaled.
       max(axis=0)))
41
42 #Train SVM model on scaled data
43 svm.fit(X_train_scaled, y_train)
44 print("Test set accuracy scaled data with Min-Max scaling: {:.2f}".
       format(svm.score(X_test_scaled,y_test)))
45 scaler = StandardScaler()
46 scaler.fit(X_train)
47 X_train_scaled = scaler.transform(X_train)
48 X_test_scaled = scaler.transform(X_test)
49
50 svm.fit(X_train_scaled, y_train)
```

```
51 print("Test set accuracy scaled data with Standar Scaler: {:.2f}".
       format(svm.score(X_test_scaled,y_test)))
52 #prediction by using scaled data
53 y_predict = svm.predict(X_test_scaled)
54 cms = np.array(confusion_matrix(y_test, y_predict, labels=[1, 0]))
55 confusion_mat_scal = pd.DataFrame(cms, index=['sick', 'healthy'],
       columns=['predictedSick', 'predictedHealthy'])
56 print(confusion_mat_scal)
57 print("___________________________________________________")
58
59 sns.heatmap(confusion_mat_scal, annot=True)
60 plt.show()
61 print(classification_report(y_test, y_predict))
```

Listing A.1: A simple Python script to classify a breast cancer dataset using the SVM algorithm available in the *scikit-learn* package.

We assume that the readers have already installed Python on their machines. For more information on setting up the Python environment for machine learning in Windows, Linux, and MacOS operating systems, readers can consult the available Python user guides related to their installed operating system. To use the Python scrip shown in Listing A.1, just copy and paste it in a Python console or using any Python editor (for instance, this code was written using PyCharm CE IDE which is freely available for academic purpose at https://www.jetbrains.com/pycharm/).

Let's dive into the code. The lines (from 1 to 10 of Listing A.1) are necessary to load into our script all the additional Python libraries necessary to import the dataset (line 6 of Listing A.1), the `sklearn` module, and the other libraries necessary to perform the data analysis. The statement at (line 11 of Listing A.1) loads the cancer dataset. Next, lines (15, 16 of Listing A.1) contain the statements necessary to split the input data set using 80% of the instances to yield the training data set and the remaining 20% to build the testing set. The statement at (line 24 of Listing A.1) defines the `y_predict` variable which is the feature that we want to predict. In this example, we would predict the type of cancer that could assume the following values "*Malignant*" or "*Benign*," whereas *X* is the predictor. To create the training and testing set, it is necessary to import `train_test_split` library, line 4. The statement at (line 21 of Listing A.1) enables us to create our model, and at line (23 of Listing A.1) we start the model's training. To evaluate the model's accuracy, we can use the *confusion matrix* (see lines 28, 29 of Listing A.1), using a *NumPy array* and *Pandas data-frame* to represent the confusion matrix. The statement at (line 30 of Listing A.1) prints on the Python console the computed confusion matrix. The next instruction visualizes on the console the confusion matrix using the `seaborn` library (let's see statement at lines 33, 34 of Listing A.1). To improve the performance of our SVM model, we must normalize the input data. The normalization process starts at line (38 of Listing A.1). In contrast, the statement at (line 50 of Listing A.1) enables us to train our SVM model using the normalized data, and, at line (61 of Listing A.1), we start the classification by using the scaled data. Finally, the remaining statements make possible verification of the improved performance in terms of the classification accuracy reached by our SVM model after training it by using normalized data.

A.2 Cluster analysis of gene expression data

This section presents a simple Python script to perform clustering analysis on gene expression data sets.

To perform clustering analysis, we will use the `gene expression cancer RNA-Seq` data set freely available from the *UCI Machine Learning Repository* https://archive.ics.uci.edu/ml/index.php at https://archive.ics.uci.edu/ml/datasets/gene+expression+cancer+RNA-Seq#. The UCI Machine Learning Repository is a collection of databases used by the machine learning community for the practical analysis of machine learning algorithms [412]. RNA sequencing (RNA-Seq) is a sequencing technique that uses next-generation sequencing (NGS) to assess the presence and abundance of RNA in a biological sample. The data set is obtained from the RNA-Seq data set [413]; it is a random collection of gene expressions of subjects presenting various types of cancer, e.g., breast carcinoma (BRCA), kidney renal clear-cell carcinoma (KIRC), colon adenocarcinoma (COAD), lung adenocarcinoma (LUAD), and prostate adenocarcinoma (PRAD).

We assume that the reader has already installed Python. For more information on setting up Python environment for machine learning in Windows, Linux, and MacOS, readers can consult the available Python user guides related to their installed operating system. To use the Python script shown in Listing A.2, just copy and paste it in a Python console or using any Python editor (for instance, this code has been written using PyCharm CE IDE, freely available for academic purpose at https://www.jetbrains.com/pycharm/).

Let's start to illustrate the main statements encompassed within the Python script presented in Listing A.2. The statements from line (1 to 10 of Listing A.2) contain the libraries import statements, necessary to support the clustering analysis.

```
1  import matplotlib.pyplot as plt
2  import tarfile
3  import urllib
4  import ssl
5  import numpy as np
6  import pandas as pd
7  import seaborn as sns
8  from sklearn.cluster import KMeans
9  from sklearn.decomposition import PCA
10 from sklearn.metrics import silhouette_score, adjusted_rand_score
11 from sklearn.preprocessing import LabelEncoder, MinMaxScaler
12 import matplotlib.pyplot as plt
```

```
13 # The PanCancer dataset url from UCI
14 uci_tcga_url = "https://archive.ics.uci.edu/ml/machine-learning-
       databases/00401/"
15 archive_name = "TCGA-PANCAN-HiSeq-801x20531.tar.gz"
16
17 #To prevent possible problem related to the SSL certificate
       verification should be used the following instruction
18 ssl._create_default_https_context = ssl._create_unverified_context
19
20 # We use the following instruction to Build the url
21 data_download_url = urllib.parse.urljoin(uci_tcga_url, archive_name)
22 #Data downloading Request and the result is saved in the variable
23 downloaded_result = urllib.request.urlretrieve(data_download_url,
       archive_name)
24
25 # unzip the TCGA data
26 tar_file = tarfile.open(archive_name, "downloaded_result:gz")
27 # store unzipped files into the selected directory dataset
28 tar.extractall(path="/Users/dataset/")
29 tar.close()#stream closed
30
31 #file containing the measured gene expression values
32 gene_expression_data_points_file_path = "/Users/dataset/TCGA-PANCAN-
       HiSeq-801x20531/data.csv"
33 #file containing the cancer labels
34 cancer_type_file_path = "/Users/dataset/TCGA-PANCAN-HiSeq-801x20531/
       labels.csv"
35
36 gene_exp_points = np.genfromtxt(gene_expression_data_points_file_path,
       delimiter=",", usecols=range(1, 20532), skip_header=1)
37 cancer_names = np.genfromtxt(cancer_type_file_path, delimiter=",",
       usecols=(1,), skip_header=1, dtype="str")
38
39 #To use these labels in the evaluation methods, it is first needed to
       convert the abbreviations to integers with LabelEncoder
40 label_encoder = LabelEncoder()
41 converted_cancer_names = label_encoder.fit_transform(cancer_names)
42
43 # to set the number of cluster K
44 num_clusters=len(label_encoder.classes_)
45 #Scale gene expression values to a given range
46 scaler = MinMaxScaler()
47 transformed_data=scaler.fit_transform(gene_exp_points)
48 #PCA instantiation and run
```

```
49 pca = PCA(n_components=2, random_state=42).fit_transform(
       transformed_data)
50
51 #kmeans instantiation
52 kmeans=KMeans(n_clusters=num_clusters, init="k-means++", n_init=50,
       max_iter=500, random_state=42).fit(pca)
53
54 #The cluster assignments are stored as a one-dimensional NumPy array
55 predicted_labels=kmeans.labels_
56
57 #K-Means performance evaluation
58 print(silhouette_score(pca, predicted_labels))
59 print(adjusted_rand_score(converted_cancer_names, predicted_labels))
60
61 dataFrame = pd.DataFrame(pca, columns=["component_1", "component_2"])
62 dataFrame["predicted_cluster"] = kmeans.labels_
63 dataFrame["converted_cancer_names"] = label_encoder.inverse_transform(
       converted_cancer_names)
64 #plt setting
65 plt.style.use("fivethirtyeight")
66 plt.figure(figsize=(8, 8))
67 scat = sns.scatterplot(
68 "component_1",
69 "component_2",
70 s=50,
71 data=dataFrame,
72 hue="predicted_cluster",
73 style="converted_cancer_names", palette="Set2")
74
75 plt.legend(bbox_to_anchor=(1.05, 1.0),loc='upper left')
76 plt.tight_layout()
77 plt.show()
```

Listing A.2: A simple script to use version of K-means algorithm available in *scikit-learn* package.

In Listing A.2 the statements from (line 1 to 11 of Listing A.2) allow the import of the external libraries needed to implement the clustering analysis. First of all, we need to download the *RNAseq* data set, a task that can be done programmatically by using the statements at (lines 13–22 of Listing A.2). The statements contained in (line 13 and 14 of Listing A.2) define the dataset *url* and the dataset *name,* respectively. To prevent possible download failures, one needs to use the statement at (line 17 of Listing A.2) which prevents possible *SSL* certificate verifications. Next, we need to build the whole url; to download the data, we can use the `urljoin` method (line 20 of Listing A.2), and at (line 22 of Listing A.2) we start the download and save the downloaded data set into the variable named `downloaded_result`. Subsequently, we need to extract the data (see the statements in lines 25–28 of Listing A.2). As the first step, it is necessary to open the `tar` stream, by using the `open` method statement at (line 25 of Listing A.2); next, we can extract the dataset on the disk by giving to the method `extractall` the path toward the selected folder (see statement at line 27 of Listing A.2), and then we can close the tar stream. As a result, on our disk locally there should be two *csv* files at the selected destination. The two files, `data.csv` and `labels.csv` contain the data and the labels, respectively. To load the csv files into the main memory, we can use the *NumPy* function `genfromtxt`. `genfromtxt` function loads the data from the csv file into memory as NumPy arrays (see statements at lines 35–36 of Listing A.2). The cancer name labels, stored with the `cancer_names` variable, have to be converted in order to make it possible to use them in the evaluation method. This conversion is mandatory because the K-means algorithm can handle only numerical values. For this purpose, we can use the `LabelEncoder` class (lines 39–40 of Listing A.2). The labelEncoder maps the cancer labels on numbers, and, through the `.classes_` keyword, it is possible to return the number of unique values used to yield the mapping. Thus, we can use this information later to set the number of clusters k for the K-means algorithm. Machine learning algorithms—K-means included—for yielding accurate and precise models require that all the features belong to the same scale. To scale all the gene expression values into a common range, we can use the `MinMaxScaler` class. The `MinMaxScaler` transforms features by scaling each feature to a given range. In particular, it scales and translates each feature individually such that it is in the given range on the training set, e.g., between zero and one. Line 45 of Listing A.2 contains the `MinMaxScaler` instantiation statement, whereas the scaled data are stored in the `transformed_data` variable (see line 46 of Listing A.2). Before running the K-means algorithm, it is advisable to reduce the number of features of the problem, known as *dimensionality reduction*. Dimensionality reduction helps to reduce the number of features since, with a high number of features, the features space becomes sparse, and the learning algorithms produce models with low accuracy and precision. To this end, we can use *Principal Component Analysis* (PCA) one of the best known and most used dimensionality reduction algorithms. At line 48 of Listing A.2, we instantiate the PCA class. At this point, we have all the data in the proper format to instantiate and run the K-means algorithm (line 51 of Listing A.2). Let's analyze the parameters used to run the K-means algorithm: `n_clusters=num_clusters` enables the definition of the k number of expected clusters. `init="k-means++"` enables definition of the initialization method; we used `"k-means++"` which speeds up the convergence, compared to the default initialization method that is random. `n_init=50` sets the number of initializations to be performed because two runs can converge on different cluster assignments. The default behavior of K-means is to perform 10 runs and return the one with the lowest SSE results, `max_iter=500` sets the maximum number of iterations, e.g., 500, and `random_state=42` defines random-number generation for centroid initialization. The statement at line 54 enables to convert the cluster assignment in a *NumPy* array. Next the array will be used to evaluate the quality of the K-means produced clusters (see statements into the lines 57–58 of Listing A.2). Quality evaluation will be done by using the `silhouette_score` and the `adjusted_rand_score` functions (available in the `sklearn.metrics` library). To evaluate the K-means cluster assignment, we can use the *silhouette* and the *Adjusted Rand Index (ARI)*. The silhouette coefficient calculates cluster coherence and separation, quantifying how a data point fits into its assigned cluster, whereas ARI measures the similarity between true and predicted labels. To visualize the produced clusters, we need to store the proper data in a *Pandas* `dataframe` by using the statements at lines 60–62 of Listing A.2. Finally, we can use `matplotlib` and `seaborn` libraries to produce a visual representation of the produced clusters (see statement at lines 64–76 of Listing A.2). Fig. A.1 is the visual representation of the produced K-means clusters.

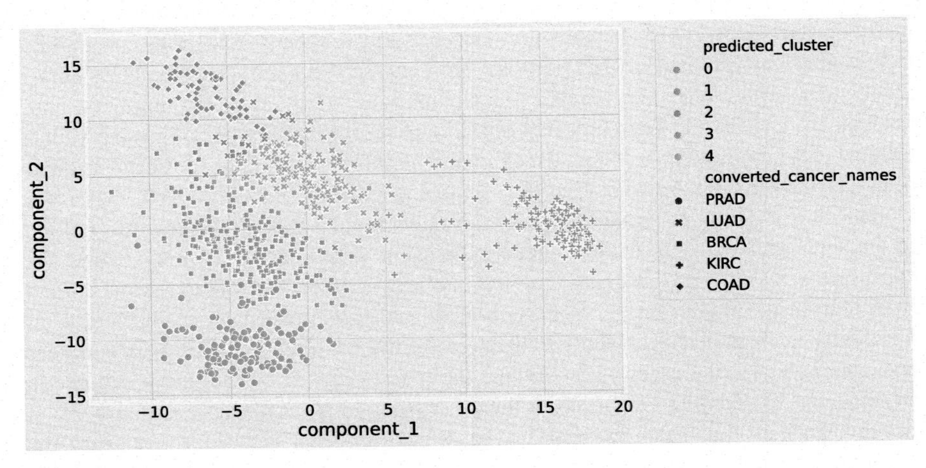

FIGURE A.1

The produced K-means clusters.

A.3 Python agent-oriented programming framework

MESA is a Python framework for the development of multi-agent models. MESA is an open-source framework released under the Apache 2.0 license. MESA allows users to create agent-based models using the built-in core components (e.g., agent schedulers and spatial grids) or specific implementations, visualizing agents using a graphical interface and exporting agents' results to be analyzed using Python's data analysis tools. MESA simplifies agent real-time visualization by using the browser as a front-end and exploiting the power of JavaScript data visualization, thus avoiding that the programmer must deal with GUI problems directly. MESA provides users a modular architecture comprising the following core elements needed to build a model: *i)* a model class to store model-level parameters providing a container for the other components; *ii)* One or more types of agent classes that specify the role of the agents; and *iii)* a scheduler module that controls the agent activation management.

To create an agent with MESA, we can use the *model* and *Agent* MESA's core classes (see Listing A.3). The model class contains the model attributes, and it manages the agents. To each model, it is possible to add multiple agents (see Listing A.3 line 27 where the agent is instantiated and line 28 agent is added to the model). All the agents are instances of the agent class. In our example, each agent has only one variable: the protein to translate (see Listing A.3 line 10). Each agent will also have a unique identifier, i.e., the protein identifier to translate (see Listing A.3 line 9) stored in the unique_id variable. Giving each agent a unique identifier is very useful, especially when dealing with multi-agent-based modeling.

```
1  package
2  from mesa import Agent, Model
3  from mesa.time import RandomActivation
4  import urllib.request
5  import ssl
6
7  class MappingAgent(Agent):
8      #An agent that receive the protein to convert.
9      def __init__(self, unique_id, model, prot):
10         super().__init__(unique_id, model)
11         self.protein = prot
12
13     def step(self):
14         print( 'Agent ', self.unique_id, "received ", 'protein ', self
       .protein)
15         ssl._create_default_https_context = ssl.
       _create_unverified_context
16         #Loading the UniProt protein to convert
17         data = urllib.request.urlopen("http://www.uniprot.org/uniprot/
       " + self.prot + ".txt").read()
18         #Print the obtain information
19         print(data)
20
21 class AgentModel(Model):
22     #A model with some agents.
23     def __init__(self, N, protein):
24         self.num_agents = N
25         self.prot = protein
26         self.schedule = RandomActivation(self)
27         # Create agents
28         a = MappingAgent(self.prot, self, pro=self.prot)
29         self.schedule.add(a)
30
```

```
31     def step(self):
32         #Advance the model by one step.
33         self.schedule.step()
34
35 proteins_to_search = [ 'P40925', 'P40926', '043175', 'Q9UM73', 'P97793
       ']
36 for i in proteins_to_search:
37     model = MoneyModel(10, i)
38     model.step()
39     model = None
40 print("end")
```

Listing A.3: A simple Mesa agent that converts UniProt identifiers.

A.4 Sequences similarity score calculation

In this section, we will cover some Python functions for calculating the similarity score of two sequences. To speed up the similarity score calculation, we can implement the *similarity matrix* as a dictionary in Python. The dictionary makes possible dealing with residues scores efficiently using keys, avoiding any worry about handling the lines and columns indexes necessary for implementing the similarity matrix like a two-dimensional array.

```
1  def read_blosum(path):
2      with open(path, 'r') as file:
3          content = file.read() #load the file content
4      lines = content.strip().split('\n') # Assign to lines a list of
       word with leading and trailing whitespace removed
5      header = lines.pop(0) # the residues list
6      columns = header.split()
7      sim_matrix = {}
8      for row in lines:
9          entries = row.split()
10         row_name = entries.pop(0)  # position 0 contains the residue
       identifier
11         sim_matrix[row_name] = {}
12         if len(entries) != len(columns):
13             raise Exception('Entry number mismatch')
14         for column_name in columns:
15             sim_matrix[row_name][column_name] = entries.pop(0)
16     return sim_matrix
17
18 def similarity(seq_a, seq_b, sim_matrix):
```

```
19    # Assign to length the minimum length between the two sequences to
        prevent index out of range exception.
20    length = min(len(seq_a), len(seq_b))
21    score = 0
22    #Similarity score is computed iterating over the two sequences
23    for i in range(length):
24      residue_a = seq_a[i]
25      residue_b = seq_b[i]
26      #the value related to the two residues is iteratively added to
        score
27      score += int(sim_matrix[residue_a][residue_b])
28    return score
29
30 def main():
31    matrix = read_blosum(path='./blosum.txt')
32    print(similarity('ACGTAGAGAG', 'CAGAGAGTGAC', sim_matrix=matrix))
33 if __name__ == "__main__":
34    main()
```

Listing A.4: Two simple Python functions to read a Blosum matrix from a file and compute a similarity score.

Listing A.4 defines two functions: the `read_blosum(path)` function (line 1 of Listing A.4) that loads and manipulates a generic *BLOSUM* matrix stored in a text file on the disk with the statement at line 3 of Listing A.4. After matrix loading matrix, it is necessary to perform some data manipulation by using the native Python string functions `strip` and `split`. The `strip`function will remove the white spaces, while the `split` function will break the single lines using the character ('*n*' - newline). As a result, the manipulated string is assigned to the variable `lines` (see Listing A.4 line 4). Next, iteratively we go through the strings contained within the `lines` variable (see Listing A.4 line 8), splitting the `row` content (Listing A.4 line 9). At line 12 of Listing A.4, the statement verifies that both variables `entries` and `columns` have the same length, in order to assign at each residue the respective value for the current row (see Listing A.4 line 15). The `similarity(`***`seq_a, seq_b, sim_matrix`***`)` computes the similarity score using the three parameters. To prevent index out of range exception, it is necessary to use the `min` function. The `min` function will assign the minimum length between the two sequences to the `len` variable (see Listing A.4 line 20). Next, the similarity score is computed iterating over the two sequences simultaneously Listing A.4 lines from 23 to 27.

A.5 Dynamic programming

The computation of the similarity of two sequences based on the method presented in Listing A.4 does not allow the correct handling of insertions and deletions (*indel*) in the two sequences under comparison. To this end, the dynamic algorithms can be used to overcome these limitations in calculating the optimal sequence alignment. Listing A.5 defines a simple Python function `N _ and _ W _ like _ func(`*seq_a*, *seq_b*, *sim_matrix*, `insert, extend)` implementing the *Needlman* and *Wunsch (N&W)* algorithm. To speed up the alignment computing, we use `numpy` arrays by importing the `numpy` module (see Listing A.5 line 1). At Listing A.5 line 2, we import the blosum reader function used in Listing A.4 and saved in a file called *blosum_reader.py*. The two statements at lines 6 and 7 of Listing A.5 are needed to store the starting values of the alignment. At lines 9 and 10 of Listing A.5, we define and initialize with the value zero two numpy arrays that will represent respectively the alignment score for each single pairing, and the path alignment containing the direction taken.

Iteratively, we go through the matrix element to compute the score taking into account the penalties (see Listing A.5 lines 12 and 13). Penalties are detected as show at lines 16 and 18 of Listing A.5, checking if the previous position (e.g., `i-1` and `j-1`) contains 1 or 2, the values used to represent the penalties. The best path is obtained by selecting the *max* among the computed path (see Listing A.5 line 24). To obtain the alignment, we iterate the path matrix going back until the *i* and *j* indices are greater than 0. At each iteration we verify which symbol adds to the alignment representation as shown at lines 37, 42, and 46 of Listing A.5. Finally, naming the method `reverse`, we obtain the correct alignment representation (see Listing A.5 lines 51 and 52).

```
1  import numpy as np
2  import blosum_reader
3
4
5  def N_and_W_like_func(seq_a, seq_b, sim_matrix, insert=8, extend=4):
6      length_seq_a = len(seq_a) + 1
7      length_seq_b = len(seq_b) + 1
8
9      score_matrix = np.array([[0] * length_seq_b for i in range(
       length_seq_a)])
10     paths_matrix = np.array([[0] * length_seq_b for i in range(
       length_seq_a)])
11
12     for i in range(1, length_seq_a):
13         for j in range(1, length_seq_b):
14             penalty_down = insert
15             penalty_horizontal = insert
16             if paths_matrix[i-1][j] == 1:
17                 penalty_down = extend
18             elif paths_matrix[i][j-1] == 2:
19                 penalty_horizontal = extend
20             similarity = int(sim_matrix[seq_a[i-1]][seq_b[j-1]])
```

```
21              paths = [score_matrix[i-1][j-1] + similarity,  # moving
    diagonally
22                       score_matrix[i-1][j-penalty_down,  # moving down
23                       score_matrix[i][j-1]-penalty_horizontal]  #
    moving horizontally
24              best_path = max(paths)
25              route = paths.index(best_path)
26              score_matrix[i][j] = best_path
27              paths_matrix[i][j] = route
28
29      align_seq_a = []
30      align_seq_b = []
31      i = length_seq_a - 1
32      j = length_seq_b - 1
33      score = score_matrix[i][j]
34
35      while i > 0 or j > 0:
36          route = paths_matrix[i][j]
37          if route == 0:  # paired residues
38              align_seq_a.append(seq_a[i - 1])
39              align_seq_b.append(seq_b[j - 1])
40              i -= 1
41              j -= 1
42          elif route == 1:  # gap in seq_b
43              align_seq_a.append(seq_a[i - 1])
44              align_seq_b.append('-')
45              i -= 1
46          elif route == 2:  # gap in seq_a
47              align_seq_a.append('-')
48              align_seq_b.append(seq_b[j - 1])
49              j -= 1
50
51      align_seq_a.reverse()
52      align_seq_b.reverse()
53      align_seq_a = ''.join(align_seq_a)
54      align_seq_b = ''.join(align_seq_b)
55      return score, align_seq_a, align_seq_b
56
57  def main():
58      score, align_a, align_b = N_and_W_like_func('ACGTCTG', '
    CCCGTAGTGCA',
59                                                          blosum_reader.
    read_blosum(path='./blosum.txt'), insert=6, extend=5)
60      print(score)
```

```
61     print(align_a)
62     print(align_b)
63 if __name__ == "__main__":
64     main()
```

Listing A.5: A simple Python functions to compute the similarity score using a method similar to Needlman and Wunch's.

A.6 Analysis of FASTQ sequences

This section will present how to perform some fundamental analysis on *FASTQ* sequences files using the `Biopython`[1] module. `Biopython` is a freely available analysis module to perform computational molecular analysis. We assume that readers have already installed both Python and Biopython on their machines. For more information on setting up Python environment in Windows, Linux, and macOS, readers should consult the available Python user guides related to the installed operating system.

Before starting the analysis, we need to download a *FASTQ* sequence file from the `International Genome Sample Resource (IGSR)`. *NGS* files, although compressed, commonly have dimensions of several gigabytes per single file. Hence, to become familiar with *NGS* data analysis methodologies, it is advisable to use small files at the beginning. In addition, to ensure the download of an *NGS* data set, even from low-speed networks, it is advisable to download from the *IGSR* website the dataset `SRR003265.filt.fastq.gz`, a relatively small dataset (28.7 MB), making it suitable for both download and analysis. To download the `SRR003265.filt.fastq.gz`, after reaching the *IGSR* website https://www.internationalgenome.org/home, paste the `SRR003265` dataset identifier into the search box located in the right-upper corner of the home page and wait for downloading.

When the download has been accomplished, we can compute and visualize the nucleotides' distribution in all the single reads. Listing A.6 presents the Python code to calculate the input file's nucleotides' distribution.

```
1 package
2 import gzip
3 from Bio import SeqIO
4 from collections import defaultdict
5 import matplotlib.pyplot as plt
6
7 """ Plot the distribution of Alleles according to its read position
    """
8 def alleles_distribution():
9     """ Open the FASTQ sequence file """
```

[1] https://biopython.org.

```
10      with gzip.open('SRR003265.filt.fastq.gz', 'rt', encoding='utf-8')
        as f_in:
11          reads = SeqIO.parse(f_in, 'fastq')
12  """ Print the first sequence """
13          read = next(reads)
14          print(read.id, read.description, read.seq, len(read))
15  """ Define the dictionaries to count the allele distribution """
16          n_cnt = defaultdict(int)
17          a_cnt = defaultdict(int)
18          c_cnt = defaultdict(int)
19          g_cnt = defaultdict(int)
20          t_cnt = defaultdict(int)
21  """ """
22          for read in reads:
23              for i, letter in enumerate(read.seq):
24                  pos = i + 1
25                  if letter == 'N':
26                      n_cnt[pos] += 1
27                  elif letter == 'A':
28                      a_cnt[pos] += 1
29                  elif letter == 'C':
30                      c_cnt[pos] += 1
31                  elif letter == 'G':
32                      g_cnt[pos] += 1
33                  elif letter == 'T':
34                      t_cnt[pos] += 1
35
36          seq_len_n = max(n_cnt.keys())
37          positions_n = range(1, seq_len_n + 1)
38          plt.plot(positions_n, [n_cnt[x] for x in positions_n], label='
        N - unknown bases - calls')
39
40          seq_len_a = max(a_cnt.keys())
41          positions_a = range(1, seq_len_a + 1)
42          plt.plot(positions_a, [a_cnt[x] for x in positions_a], label='
        A - Adenine- calls')
43
44          seq_len_c = max(c_cnt.keys())
45          positions_c = range(1, seq_len_c + 1)
46          plt.plot(positions_c, [c_cnt[x] for x in positions_c], label='
        C - Cytosine - calls')
47
48          seq_len_g = max(g_cnt.keys())
49          positions_g = range(1, seq_len_g + 1)
```

```
50         plt.plot(positions_g, [g_cnt[x] for x in positions_g], label='
        G - Guanine - calls')
51
52         seq_len_t = max(t_cnt.keys())
53         positions_t = range(1, seq_len_t + 1)
54         plt.plot(positions_t, [t_cnt[x] for x in positions_t], label='
        T - Timine - calls')
55
56         plt.legend(loc='lower left')
57         plt.show()
58
59 def main():
60     alleles_distribution()
61
62 if __name__ == '__main__':
63     main()
```

Listing A.6: A Python function exploiting Biopython to compute the alleles distribution with regards to the spatial distribution.

The downloaded dataset has a size that makes possible it being loaded entirely into central memory (this is difficult to achieve with a *FastaQ* file of a few gigabytes). Hence, we can load and unpack it using `gzip` (see Listing A.6 line 9) as a first operation, and we can also specify that we approaching to load and unpack a *FastaQ* file (through the keyword 'fastq', see Listing A.6 line 10). Recall importing `gzip` (see Listing A.6 line 1) to make it available in the program. Although decompressing is a slow process, it can still be faster than reading a much bigger (uncompressed) file from a disk. At Listing A.6 line 10, we use `Seq.IO.parse` class to load the input file. As a rule, when you want to inspect a file, it is better always to iterate over the file than to list all the content using `list` command, especially with huge files like *NGS,* with the result of overwhelming the main memory. If you want just to see the content in the input file, you can print the content of a single line and print it. To read the content of the current line you must use the statement at Listing A.6 line 13. Next, to print the contents of the line just read you must print it, by using the statement at Listing A.6 line 14. Instead, if you have to perform several reading operations over it (this also applies to write operations), a better way to obtain the data from the file is to iterate cyclically over it, as done in Listing A.6 line 21, 21, and get the current read. A single read contains several types of information implemented as an enumeration using the keyword `enumerate` (see Listing A.6 line 22). We can use a `for` cycle to iterate over the enumeration, to retrieve the alleles identifier, and to increment the allele's counter at that position (see Listing A.6 line 23), storing it in the respective dictionary (see Listing A.6 lines from 24 up to 33). Then, on each dictionary we perform some operation helpful to plot for each allele its distribution as a function of the distance from the start of the sequence (see Listing A.6 lines from 35 to 53). Finally, the instruction at line 55 adds the legend to the cart, whereas instruction at line 56 visualizes the chart (see Fig. A.2).

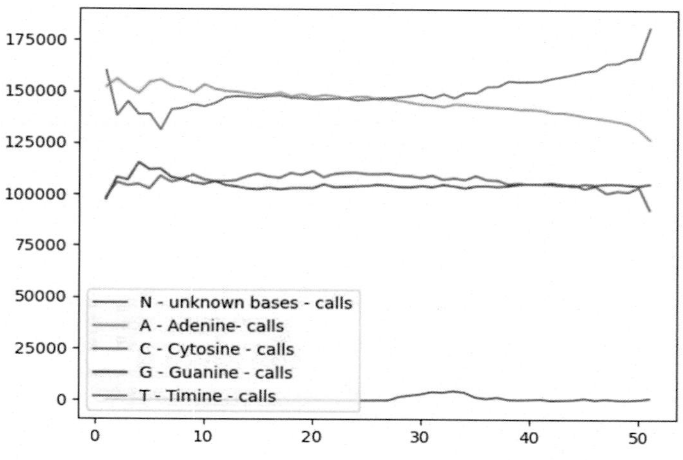

FIGURE A.2

The number of alleles calls as a function of the distance from the start of the sequencer read.

A.7 Analysis of alignment map in SAM/BAM format

In this section, we will see how to perform some fundamental analysis on *SAM/BAM* alignment files using the *Pysam*[2] module. *Pysam* is a freely available analysis module for reading, manipulating, and writing *SAM/BAM* files. We assume that readers have already installed Python and Pysam in their machines. For more information on setting up Python environment in Windows, Linux, and macOS operating systems, the reader should consult the available Python user guides related to the installed operating system.

Before performing some fundamental analysis using *SAM/BAM* file, it is advisable to download *SAM/BAM* files from the *ENCODE* web-directory.[3] *ENCODE* contains a collection of alignment files starting from around 40 MB that are more suitable for our analysis purposes. This is a good starting point to become familiar with the alignment analysis, considering that the average magnitude of real *SAM/BAM* files is on the order of some GBytes. As an example, the whole human genome alignment magnitude is about 40 GB.

To perform the analysis, we need to download two files: the `wgEncodeUwRepliSeqBjG1bAlnRep1.bam` with dimension of about 62.5 MB and `wgEncodeUwRepliSeqBjG1bAlnRep1.bam.bai` with a size of about 5.4 MB. To download the data sets, we can access the ENCODE website http://hgdownload.cse.ucsc.edu/goldenPath/hg19/encodeDCC/wgEncodeUwRepliSeq, selecting the two files to download, or using

[2] https://pypi.org/project/pysam/.

[3] http://hgdownload.cse.ucsc.edu/goldenPath/hg19/encodeDCC/wgEncodeUwRepliSeq.

the `curl` command. To use *curl*, open a command line shell and type the following command:
`curl` http://hgdownload.cse.ucsc.edu/goldenPath/hg19/encodeDCC/wgEncodeUwRepliSeq/wgEncodeUwRepliSeqBjG1bAlnRep1.bam > `wgEncodeUwRepliSeqBjG1bAlnRep1.bam`; then, press return and wait until the download is completed, and repeat for the second file `curl` http://hgdownload.cse.ucsc.edu/goldenPath/hg19/encodeDCC/wgEncodeUwRepliSeq/wgEncodeUwRepliSeqBjG1bAlnRep1.bam.bai > `wgEncodeUwRepliSeqBjG1bAlnRep1.bam.bai`.

Let's analyze the code shown in Listing A.7.

```
1  package
2  import pysam
3
4  bam = pysam.AlignmentFile('/Users/giuseppeagapito/PycharmProjects/
       pythonProject4BookAI/wgEncodeUwRepliSeqBjG1bAlnRep1.bam','rb')
5  headers = bam.header
6  for seq_type, sequences in headers.items():
7      print(seq_type)
8      for i, seq in enumerate(sequences):
9          if type(seq) == dict:
10             print("\t%d" % (i+1))
11             for k, v in seq.items():
12                 print("\t\t%s\t%s" % (k, v))
13         else:
14             print("\t\t%s" % seq)
15 for sequence in bam:
16     if(not sequence.is_unmapped and not sequence.mate_is_unmapped):
17         break
18 print(sequence.query_name, sequence.reference_id, bam.getrname(
       sequence.reference_id), sequence.reference_start, sequence.
       reference_end)
19 print(sequence.cigarstring, '1')
20 print(sequence.query_alignment_start, sequence.query_alignment_end,
       sequence.query_alignment_length, 'k')
21 print(sequence.next_reference_id, sequence.next_reference_start,
       sequence.template_length, 'h')
22 print(sequence.is_paired, sequence.is_proper_pair, sequence.
       is_unmapped, sequence.mapping_quality)
23 print(sequence.query_sequence, '?')
```

Listing A.7: A Python function exploiting pySam to perform SAM/BAM alignment.

First, we import the `pysam` module to make it available for use in the program. Next, we use `AlignmentFile()` method to create an alignment-file object and store it in a variable named `bam`. Once the file is open, it is possible to iterate its content, e.g., the read mappings. Each iteration returns an `AlignedSegment` object. Calling `header` on `bam` variable, e.g., `headers = bam.header` (see Listing A.7 line 4), we obtain a multi-level dictionary. Using `headers.items()` enables iterating over all the content of the dictionary and printing the sequences using as the formatting operator the character '`%`'. The formatting operator's syntax is (*format % values*); the % operator replaces the specifications in format with zero or more elements of values. For instance, the statement "(\t%s\t%s" % (k, v)) means that, %s%s are strings placeholder, and their value will be mapped in the tuple (k, v), whereas *%d* is a placeholder for numbers. We iterate over the `bam` variable until we find the first mapped sequence, visualizing some basic sequence properties. To obtain indication about the alignment of the single base, we can use *CIGAR* [414] class (Compact Idiosyncratic Gapped Alignment Report) that represents a compressed (run-length encoded) pairwise alignment format. Fig. A.3 shows the results obtained using the CIGAR alignment interrogation.

```
    20
        SN   chr20
        LN   63025520
    21
        SN   chr21
        LN   48129895
    22
        SN   chr22
        LN   51304566
    23
        SN   chrM
        LN   16571
    24
        SN   chrX
        LN   155270560
SOLEXA-1GA-1_2_FC20ENP:5:91:381:65 0 chr1 10157 10184
27M l
0 27 27 k
-1 -1 0 h
False False False 25
AACCCTAACCCTAACCCTAACCTAACC ?

Process finished with exit code 0
```

FIGURE A.3

The results obtained by using the Pysam alignment module.

A.8 Mass spectrometer data analysis

Mass spectrometer output files contain the mass spectra resulting from MS runs. MS output files may be broadly classified into three subclasses: proprietary vendor, complex open, and simple text formats. The mass spectra data are classified in profile-mode (also named continuous) spectra and as peak lists (also named centroided or peak-picked), and they are available in several different formats:

- raw format contains profile mode spectra or centroided spectra depending on the user's selection. Raw output is available in the formats mzML, mzXML, netCDF, and mzData.
- mzIdentML format is an exchange standard for peptide and protein identification data, designed by the Proteomics Standards Initiative.
- mzQuantML format extends the mzIdentML format, including the ability to encode quantification information.
- MGF format encodes multiple MS/MS spectra into a single file.

Fig. A.4 shows an example of a mzML file compliant with the standard developed by HUPO.

Let's see how to open and perform some basic operation on `mzML` files using `pyOpenMS` Python module.

First, we need a `mzML` file that we can download from the *HUPO* website at the following link, http://proteowizard.sourceforge.net/example_data/tiny.pwiz.1.1.mzML. Listing A.8 shows a simple code to perform basic data filtering on spectra data that have values under a given threshold. To avoid to manually investigate the spectra, we can automatically set the threshold value as the minimum spectra value. Note that it is possible to replace the threshold setting criteria (see Listing A.8 line 6) with the average, the mode, or the weighted average. The method `setup_threshold(spectra)` (see Listing A.8 line 21) compares all the spectra to figure out the minimum and returns it.

```
1  package
2  from pyopenms import *
3  def ms_filter(spectra, threshold):
4      count_discarded = 0
5      exp = MSExperiment()
6      for spec in spectra:
7          if spec.getMSLevel() > threshold:
8              exp.addSpectrum(spec)
9          else:
10             count_discarded += 1
11     print 'discarded spectra: ', count_discarded
12     MzMLFile().store("./filteredMS.mzML", exp)
13 def setup_threshold(spectra):
14     min = 1000
15     for spec in spectra:
16         if spec.getMSLevel() < min:
17             min = spec.getMSLevel()
18     return min
19 def main():
20     spectra = MSExperiment()
21     MzMLFile().load("./tiny.pwiz.1.1.mzML", spectra)
```

```
22     ms_filter(spectra=spectra, threshold=setup_threshold(spectra=
       spectra))
23 if __name__ == '__main__':
24     main()
```

Listing A.8: A mass spectra filtering script.

To filter all the spectra under the given threshold, we defined the `ms_filter(spectra, threshold)` method. The method compares all the spectra level and filters out all the spectra whose values are less than the threshold (see Listing A.8, line 8). Finally, all the selected spectra will be stored in a new `mzML` file in the specified folder (see Listing A.8 line 20); in this example the output file will be saved in the current directory.

```
<?xml version="1.0" encoding="ISO-8859-1"?>
<indexedmzML xmlns="http://psi.hupo.org/ms/mzml"
xmlns:xsi="http://www.w3.org/2001/XMLSchema-instance"
xsi:schemaLocation="http://psi.hupo.org/ms/mzml
http://psidev.info/files/ms/mzML/xsd/mzML1.1.0_idx.xsd">
  <mzML xmlns="http://psi.hupo.org/ms/mzml"
xmlns:xsi="http://www.w3.org/2001/XMLSchema-instance"
xsi:schemaLocation="http://psi.hupo.org/ms/mzml
http://psidev.info/files/ms/mzML/xsd/mzML1.1.0.xsd"
id="urn:lsid:psidev.info:mzML.instanceDocuments.tiny.pwiz" version="1.1.0">
    <cvList count="2">
      <cv id="MS" fullName="Proteomics Standards Initiative Mass Spectrometry Ontology"
version="2.26.0"
URI="http://psidev.cvs.sourceforge.net/*checkout*/psidev/psi/psi-ms/mzML/
controlledVocabulary/psi-ms.obo"/>
      <cv id="UO" fullName="Unit Ontology" version="14:07:2009"
URI="http://obo.cvs.sourceforge.net/*checkout*/obo/obo/ontology/phenotype/unit.obo"/>
    </cvList>
    <fileDescription>
      <fileContent>
        <cvParam cvRef="MS" accession="MS:1000580" name="MSn spectrum" value=""/>
        <cvParam cvRef="MS" accession="MS:1000127" name="centroid spectrum" value=""/>
      </fileContent>
      <sourceFileList count="3">
        <sourceFile id="tiny1.yep" name="tiny1.yep" location="file://F:/data/Exp01">
          <cvParam cvRef="MS" accession="MS:1000567" name="Bruker/Agilent YEP file"
value=""/>
          <cvParam cvRef="MS" accession="MS:1000569" name="SHA-1"
value="1234567890123456789012345678901234567890"/>
          <cvParam cvRef="MS" accession="MS:1000771" name="Bruker/Agilent YEP nativeID
format" value=""/>
        </sourceFile>
        <sourceFile id="tiny.wiff" name="tiny.wiff" location="file://F:/data/Exp01">
          <cvParam cvRef="MS" accession="MS:1000562" name="ABI WIFF file" value=""/>
          <cvParam cvRef="MS" accession="MS:1000569" name="SHA-1"
value="2345678901234567890123456789012345678901"/>
          <cvParam cvRef="MS" accession="MS:1000770" name="WIFF nativeID format"
value=""/>
        </sourceFile>
        <sourceFile id="sf_parameters" name="parameters.par"
location="file://C:/settings/">
          <cvParam cvRef="MS" accession="MS:1000740" name="parameter file" value=""/>
          <cvParam cvRef="MS" accession="MS:1000569" name="SHA-1"
value="3456789012345678901234567890123456789012"/>
          <cvParam cvRef="MS" accession="MS:1000824" name="no nativeID format" value=""/>
        </sourceFile>
      </sourceFileList>
      <contact>
        <cvParam cvRef="MS" accession="MS:1000586" name="contact name" value="William
Pennington"/>
        <cvParam cvRef="MS" accession="MS:1000590" name="contact organization"
value="Higglesworth University"/>
        <cvParam cvRef="MS" accession="MS:1000587" name="contact address" value="12
Higglesworth Avenue, 12045, HI, USA"/>
        <cvParam cvRef="MS" accession="MS:1000588" name="contact URL"
```

FIGURE A.4

An example of an mzML file.

APPENDIX

Java code examples

B

Throughout this book, we shall use Java to perform some omics data analysis, employing the most suitable third-party intelligent agent packages, namely, *JADE* and *JADEX*. The goal of this appendix is to provide a basic understanding and skills in implementing intelligent software agents in *JADE* and *JADEX*.

B.1 Java agent-oriented programming frameworks

B.1.1 Java Agent DEvelopment (JADE) framework

JADE is an agent-oriented programming (AOP) framework implemented in Java. It simplifies the implementation of Multi-Agents Systems (*MAS*) through a middleware that complies with *FIPA* specifications. *FIPA* is an *IEEE* Computer Society standards organization that promotes agent-based technology and its interoperability standards with other technologies. Moreover, *JADE* provides graphical tools supporting the debugging and deployment phases. Distributed *JADE* agents can operate across machines (even with different operating systems), allowing the configuration of distributed agents via Graphical User Interface (*GUI*). In addition, *JADE* provides a simple yet powerful task execution and composition model, a synchronous peer-to-peer message-passing paradigm agent communication, a yellow-pages service helping publish–subscribe discovery mechanism, and many other advanced features for the development of a distributed system. JADE is free software distributed by Telecom Italia as open-source under the LGPL's terms and conditions (Lesser General Public License Version 2) license at https://jade.tilab.com.

```
1 package test.agent;
2 import jade.core.Agent;
3
4 public class SimpleJadeAgent extends Agent{
5     protected void setup() {
6                 System.out.println("Ciao World! My name is Jade Agent
     007");
7                 try {
8                         UniprotDownloader uniDown = new
     UniprotDownloader();
```

```
9                       uniDown.run("uploadlists", new
    ParameterNameValue[] {
10                                      new ParameterNameValue("from",
     "ACC"),
11                                      new ParameterNameValue("to", "
    P_REFSEQ_AC"),
12                                      new ParameterNameValue("format
    ", "tab"),
13                                      new ParameterNameValue("query"
    , "P13368 P20806 Q9UM73 P97793 Q17192"),
14                      });
15              }catch(Exception e) {}
16
17              public static void main(String[] args) {
18              SimpleAgent a = new SimpleAgent();
19              a.setup();
20      }
21          }}//end SimpleJadeAgentClass
```

Listing B.1: A simple JADE Agent to convert a list of *UniProt* identifiers to their respective *RefSeq* Protein identifiers.

Once *JADE* has been downloaded, it is necessary to include the `jade.jar` file in your system classpath or import it as an additional library into your chosen *Java IDE* for code development. We suppose that the `jade.jar` file has been already added properly to the system classpath or into the preferred IDEs. More information on using *JADE* is available at https://jade.tilab.com/doc/tutorials/JADEProgramming-Tutorial-for-beginners.pdf. Agent creation in *JADE* is simple. The new implemented class must extend the class `agent` available in the package "`jade.core`" (see Listing B.1 line 2) and implement the method "`setup()`". The *setup()* method is intended to include agent initialization, and we also use it to perform the *Uniprot* identifier mapping in the respective *RefSeq* Protein identifiers, by means of the `UniProtDownloader` class (see Listing B.1 line 8). To perform the mapping, we use the `run()` method (see Listing B.1 line 9) available in the `UniProtDownloader` class (see Listing B.2).

```
1 package test.agent;
2 import java.io.InputStream;
3 import java.io.UnsupportedEncodingException;
4 import java.net.HttpURLConnection;
5 import java.net.URL;
6 import java.net.URLConnection;
7 import java.net.URLEncoder;
8 import java.util.logging.Logger;
9
10 public class UniprotDownloader {
```

```
11      private static final String UNIPROT_SERVER = "https://www.
    uniprot.org/";
12      private static final Logger LOG = Logger.getAnonymousLogger();
13
14      public void run(String tool,ParameterNameValue[] params)
    throws Exception {
15              StringBuilder locationBuilder = new StringBuilder(
    UNIPROT_SERVER + tool + "/?");
16              for (int i = 0; i < params.length; i++)
17              {
18                      if (i > 0)
19                              locationBuilder.append('&');
20                      locationBuilder.append(params[i].name).append(
    '=').append(params[i].value);
21              }
22              String location = locationBuilder.toString();
23              URL url = new URL(location);
24              LOG.info("Submitting...");
25              HttpURLConnection conn = (HttpURLConnection) url.
    openConnection();
26              HttpURLConnection.setFollowRedirects(true);
27              conn.setDoInput(true);
28              conn.connect();
29
30              int status = conn.getResponseCode();
31              while (true)
32              {
33                      int wait = 0;
34                      String header = conn.getHeaderField("Retry-
    After");
35                      if (header != null)
36                              wait = Integer.valueOf(header);
37                      if (wait == 0)
38                              break;
39                      LOG.info("Waiting (" + wait + ")...");
40                      conn.disconnect();
41                      Thread.sleep(wait * 1000);
42                      conn = (HttpURLConnection) new URL(location).
    openConnection();
43                      conn.setDoInput(true);
44                      conn.connect();
45                      status = conn.getResponseCode();
46              }
47              if (status == HttpURLConnection.HTTP_OK)
```

```
48              {
49                      LOG.info("Got a OK reply");
50                      InputStream reader = conn.getInputStream();
51                      URLConnection.guessContentTypeFromStream(
    reader);
52                      StringBuilder builder = new StringBuilder();
53                      int a = 0;
54                      while ((a = reader.read()) != -1)
55                      {
56                              builder.append((char) a);
57                      }
58                      System.out.println(builder.toString());
59              }
60              else
61                      LOG.severe("Failed, got " + conn.
    getResponseMessage() + " for "
62                                      + location);
63              conn.disconnect();
64      }
65
66      public static class ParameterNameValue
67      {
68              private final String name;
69              private final String value;
70
71              public ParameterNameValue(String name, String value)
72                              throws UnsupportedEncodingException
73              {
74                      this.name = URLEncoder.encode(name, "UTF-8");
75                      this.value = URLEncoder.encode(value, "UTF-8")
    ;
76              }
77      }
78 }
```

Listing B.2: The UniprotDownloader class to support the agent in the conversion of Uniprot identifiers.

The `run()` method requires two parameters: The first one is the type of operation to perform, e.g., "`uploadlists`" in this case; the second one is the properties list (see Listing B.2 lines to 9 up 14). We implemented the agent's task in the `setup()` method since this is a simple example and typically the agents' task must be defined in the "`Behavior`" class (see Listing B.3).

```
1 package test.agent;
2 import jade.core.Agent;
```

```
3
4 public class SimpleJadeAgentBehaviour extends Behaviour {
5   public void action() {
6     while (true) {
7       // do something
8 } }
9   public boolean done() {
10     return true;
11 } }//endBehaviour
12 public static class SimpleJadeAgent extends Agent{
13
14 //...
15 // Add the behaviour to the agents
16     addBehaviour(new SimpleJadeAgentBehaviour());
17 }//endMyAgent
```

Listing B.3: An example of the Jade Behavior class.

As an exercise, the reader can move the *UniProt* mapping from the method `setup()` to the `action()` method and add the behavior to the `SimpleJadeAgent` class.

B.1.2 JADEX Java Agent DEvelopment eXtension framework

JADEX is an agent-oriented reasoning engine for writing rational agents with *XML* and the *Java* programming language. In *JADEX*, rational agents implement the well-known *Belief, Desire, Intention (BDI)* paradigm. *JADEX* facilitates the *BDI* use by introducing beliefs, goals, and plans as objects that can be created and manipulated inside the agent. The agents' beliefs, goals, and plans are represented as XML files. The plans enable agents to achieve their goals by executing plans that are Java methods. Agents in *JADEX* are processes consisting of two elements. The first element enables the agent to react to incoming messages, internal events, and goals by selecting and executing plans. The second one empowers the agent to estimate its current goals in order to determine a consistent strategy. In the latest version, *JADEX* allows the development of distributed applications using *Active Components*. The active components approach combines a hierarchical service component architecture (*SCA*) providing a communication model that is service-based. *JADEX* is an open-source software (*OSS*) under *GPL-License Version 3*, available at https://www.activecomponents.org/index.html#/project/news.

```
1 package
2 package agents;
3
4 import jadex.base.IPlatformConfiguration;
5 import jadex.base.PlatformConfigurationHandler;
6 import jadex.base.Starter;
7 import jadex.bridge.service.annotation.OnStart;
```

```
8 import jadex.commons.future.IFuture;
9 import jadex.micro.annotation.Agent;
10 @Agent
11 public class UniprotMapperJadexAgent{
12       @OnStart
13       public IFuture<Void> mapping(){
14             System.out.println("Ciao I'm a JadeX agent!");
15             UniprotDownloader uniDown = new UniprotDownloader();
16             uniDown.run("uploadlists", new ParameterNameValue[] {
17                         new ParameterNameValue("from", "ACC"),
18                         new ParameterNameValue("to", "
    P_REFSEQ_AC"),
19                         new ParameterNameValue("format", "tab"
    ),
20                         new ParameterNameValue("query", "
    P13368 P20806 Q9UM73 P97793 Q17192"),
21             });
22             return IFuture.DONE;
23       }
24       public static void main(String[] args){
25             IPlatformConfiguration configuration =
    PlatformConfigurationHandler.getDefaultNoGui();
26             configuration.addComponent(HelloAgent.class);
27             Starter.createPlatform(configuration).get();
28       }
29 }
```

Listing B.4: A simple JADEX agent that converts a UniProt identifier.

To use the standalone version of *JADEX*, it is necessary to set up properly the *JADEX* environment, by setting the Java *classpath*. As the first step, the *JADEX* distribution (which is the downloaded jar file) should be extracted to some local directory, for example, called `JADEX_HOME`. Next, import all the jars contained in the JADEX_HOME/lib directory into the IDE's project. JADEX agents are created using the available JADEX annotations, e.g., '`@Agent`' that identifies a class as JADEX Agent, `@AgentBody` designs a method as the agent's body and `@OnStart` designs a method to call upon start of the element; the entire JADEX annotations are available at https://www.activecomponents.org/index.html#/docs/overview. To identify a class as an agent that can be started on the platform, it is necessary to use the "`@Agent`" annotation. In addition, to speed up the scanning for running agents, all the agent classes must end with '`Agent`' by convention. Before the '`Agent`' part, any valid Java identifier can be used, for instance, `UniprotMapperJADEXAgent` (see Listing B.4 line 10). In order to convert the proteins identifier list using the `UniprotDownloader` class, we can annotate the `mapping()` method as the method to call after that the JADEX environment has been started (see Listing B.4 lines 11 and 12). At Listing B.4 line 24 is instantiated a default environment without graphical interface, while at line 25 of Listing B.4 we add the agent to the environment and, finally, we start the agent Listing B.4 line 26.

Bibliography

[1] R.J. Brachman, H.J. Levesque, Knowledge Representation and Reasoning, Morgan Kaufmann Series in Artificial Intelligence, ISBN 9781558609327, 2004, https://www.elsevier.com/books/knowledge-representation-and-reasoning/brachman/978-1-55860-932-7.

[2] S.C. Shapiro, Knowledge Representation and Reasoning: Logics for Artificial Intelligence, Technical Report, University at Buffalo, The State University of New York, Buffalo, USA, 2008.

[3] C. Baral, G. De Giacomo, Knowledge representation and reasoning: What's hot, in: B. Bonet, S. Koenig (Eds.), Proceedings of the Twenty-Ninth AAAI Conference on Artificial Intelligence, January 25-30, 2015, Austin, Texas, USA, AAAI Press, 2015, pp. 4316–4317, http://www.aaai.org/ocs/index.php/AAAI/AAAI15/paper/view/10017.

[4] E.N. Zalta, Stanford Encyclopedia of Philosophy, Stanford University, 2021, https://plato.stanford.edu/info.htmle.

[5] J. McCarthy, Programs with common sense, in: Proceedings of the Teddington Conference on the Mechanization of Thought Processes, Her Majesty's Stationary Office, London, 1959, pp. 75–91, http://www-formal.stanford.edu/jmc/mcc59.html.

[6] B.C. Smith, Reflection and Semantics in a Procedural Language, Ph.D. Thesis and Tech. Report MIT/LCS/TR-272, MIT, Cambridge, MA, 1982, http://publications.csail.mit.edu/lcs/pubs/pdf/MIT-LCS-TR-272.pdf.

[7] A.M. Turing, Computing machinery and intelligence, Mind 59 (1950) 433–460, https://doi.org/10.1093/mind/LIX.236.433.

[8] R. Epstein, G. Roberts, G. Beber, Parsing the Turing Test, Springer, ISBN 978-1-4020-6708-2, 2009, https://doi.org/10.1007/978-1-4020-6710-5.

[9] V. Singh, A. Kumar, Advances in Bioinformatics, Springer, ISBN 978-981-33-6190-4, 2021, https://doi.org/10.1007/978-981-33-6191-1.

[10] G. Agapito, M. Settino, F. Scionti, E. Altomare, P.H. Guzzi, P. Tassone, et al., DMET genotyping: tools for biomarkers discovery in the era of precision medicine, High-Throughput 9 (2) (2020), https://doi.org/10.3390/ht9020008, https://www.mdpi.com/2571-5135/9/2/8.

[11] P.H. Guzzi, G. Agapito, M. Milano, M. Cannataro, Methodologies and experimental platforms for generating and analysing microarray and mass spectrometry-based omics data to support P4 medicine, Briefings in Bioinformatics 17 (4) (2016) 553–561, https://doi.org/10.1093/bib/bbv076.

[12] M. Cannataro, G. Agapito, Computing for bioinformatics, in: S. Ranganathan, M. Gribskov, K. Nakai, C. Schönbach (Eds.), Encyclopedia of Bioinformatics and Computational Biology - Volume 1, Elsevier, 2019, pp. 160–175, https://doi.org/10.1016/b978-0-12-809633-8.20363-.

[13] H.P. Sánchez, A. Fassihi, J.M. Cecilia, H.H. Ali, M. Cannataro, Applications of high performance computing in bioinformatics, computational biology and computational chemistry, in: F.M.O. Guzman, I. Rojas (Eds.), Bioinformatics and Biomedical Engineering - Third International Conference, IWBBIO 2015. Proceedings, Part II, Granada, Spain, April 15–17, 2015, in: Lecture Notes in Computer Science, vol. 9044, Springer, 2015, pp. 527–541, https://doi.org/10.1007/978-3-319-16480-9_51.

[14] M. Settino, M. Cannataro, MMRFBiolinks: an R-package for integrating and analyzing MMRF-CoMMpass data, Briefings in Bioinformatics 22 (5) (2021), https://doi.org/10.1093/bib/bbab050.

[15] G. Agapito, M. Cannataro, cPEA: a parallel method to perform pathway enrichment analysis using multiple pathways databases, Soft Computing 24 (2020) 17561–17572, https://doi.org/10.1007/s00500-020-05243-6.

[16] G. Agapito, B. Calabrese, P.H. Guzzi, G. Fragomeni, G. Tradigo, P. Veltri, et al., Parallel and cloud-based analysis of omics data: Modelling and simulation in medicine, in: I.V. Kotenko, Y. Cotronis, M. Daneshtalab (Eds.), 25th Euromicro International Conference on Parallel, Distributed and Network-based Processing, PDP 2017, St. Petersburg, Russia, March 6-8, 2017, IEEE Computer Society, 2017, pp. 519–526, https://doi.org/10.1109/PDP.2017.68.

[17] B. Calabrese, M. Cannataro, Cloud computing in healthcare and biomedicine, Scalable Computing: Practice and Experience 16 (1) (2015), http://www.scpe.org/index.php/scpe/article/view/1057.

[18] S. Shalev-Shwartz, S. Ben-David, Understanding Machine Learning: From Theory to Algorithms, Cambridge University Press, Cambridge, UK, 2014, https://doi.org/10.1017/CBO9781107298019.

[19] M.I. Jordan, T.M. Mitchell, Machine learning: Trends, perspectives, and prospects, Science 349 (6245) (2015) 255–260.

[20] C. Donalek, Supervised and unsupervised learning, in: Astronomy Colloquia. USA, vol. 27, 2011.

[21] T.A. Kumbhare, S.V. Chobe, An overview of association rule mining algorithms, International Journal of Computer Science and Information Technologies 5 (1) (2014) 927–930.

[22] R.S. Sutton, A.G. Barto, Reinforcement Learning: An Introduction, MIT Press, 2018.

[23] T. Hastie, R. Tibshirani, J. Friedman, Overview of supervised learning, in: The Elements of Statistical Learning, Springer, 2009, pp. 9–41.

[24] S.A. Kumar, H. Kumar, S.R. Swarna, V. Dutt, Early diagnosis and prediction of recurrent cancer occurrence in a patient using machine learning, European Journal of Molecular & Clinical Medicine 7 (7) (2020) 6785–6794.

[25] H.B. Barlow, Unsupervised learning, Neural Computation 1 (3) (1989) 295–311.

[26] T.S. Madhulatha, An overview on clustering methods, arXiv preprint, arXiv:1205.1117, 2012.

[27] P.E. Danielsson, Euclidean distance mapping, Computer Graphics and Image Processing 14 (3) (1980) 227–248.

[28] F.E. Szabo, M, in: F.E. Szabo (Ed.), The Linear Algebra Survival Guide, Academic Press, Boston, ISBN 978-0-12-409520-5, 2015, pp. 219–233, https://doi.org/10.1016/B978-0-12-409520-5.50020-5, https://www.sciencedirect.com/science/article/pii/B9780124095205500205.

[29] S.B. Kotsiantis, I. Zaharakis, P. Pintelas, et al., Supervised machine learning: A review of classification techniques, Emerging Artificial Intelligence Applications in Computer Engineering 160 (1) (2007) 3–24.

[30] V. Krishnaiah, G. Narsimha, N.S. Chandra, Diagnosis of lung cancer prediction system using data mining classification techniques, International Journal of Computer Science and Information Technologies 4 (1) (2013) 39–45.

[31] S. Chatterjee, A.S. Hadi, Regression Analysis by Example, John Wiley & Sons, 2013.

[32] D.E. Rumelhart, G.E. Hinton, R.J. Williams, Learning internal representations by error propagation, Tech. rep., California Univ San Diego La Jolla Inst for Cognitive Science, 1985.

[33] F. Pedregosa, G. Varoquaux, A. Gramfort, V. Michel, B. Thirion, O. Grisel, et al., Scikit-learn: Machine learning in Python, Journal of Machine Learning Research 12 (2011) 2825–2830.

[34] L. Buitinck, G. Louppe, M. Blondel, F. Pedregosa, A. Mueller, O. Grisel, et al., API design for machine learning software: experiences from the scikit-learn project, in: ECML PKDD Workshop: Languages for Data Mining and Machine Learning, 2013, pp. 108–122.

[35] R. Agrawal, R. Srikant, et al., Fast algorithms for mining association rules, in: Proc. 20th Int. Conf. Very Large Data Bases, VLDB, vol. 1215, Citeseer, 1994, pp. 487–499.

[36] C. Borgelt, R. Kruse, Induction of association rules: Apriori implementation, in: Compstat, Springer, 2002, pp. 395–400.

[37] J. Li, L. Zhang, H. Zhou, M. Stoneking, K. Tang, Global patterns of genetic diversity and signals of natural selection for human ADME genes, Human Molecular Genetics 20 (3) (2011) 528–540, https://doi.org/10.1093/hmg/ddq498.

[38] G. Agapito, P.H. Guzzi, M. Cannataro, DMET-Miner: Efficient discovery of association rules from pharmacogenomic data, Journal of Biomedical Informatics 56 (2015) 273–283.

[39] P. Guzzi, G. Agapito, M. Di Martino, M. Arbitrio, P. Tassone, P. Tagliaferri, et al., DMET-Analyzer: automatic analysis of Affymetrix DMET data, BMC Bioinformatics 13 (1) (2012) 258, https://doi.org/10.1186/1471-2105-13-258, http://www.biomedcentral.com/1471-2105/13/258.

[40] G. Agapito, P.H. Guzzi, M. Cannataro, Parallel extraction of association rules from genomics data, Applied Mathematics and Computation 350 (2019) 434–446, https://doi.org/10.1016/j.amc.2017.09.026, https://www.sciencedirect.com/science/article/pii/S0096300317306471.

[41] G. Agapito, P.H. Guzzi, M. Cannataro, Parallel and distributed association rule mining in life science: A novel parallel algorithm to mine genomics data, Information Sciences (2021), https://doi.org/10.1016/j.ins.2018.07.055, https://www.sciencedirect.com/science/article/pii/S0020025518305723.

[42] L.P. Kaelbling, M.L. Littman, A.W. Moore, Reinforcement learning: A survey, Journal of Artificial Intelligence Research 4 (1996) 237–285.

[43] G. Lample, D.S. Chaplot, Playing FPS games with deep reinforcement learning, in: Thirty-First AAAI Conference on Artificial Intelligence, 2017.

[44] J. Kober, J.A. Bagnell, J. Peters, Reinforcement learning in robotics: A survey, The International Journal of Robotics Research 32 (11) (2013) 1238–1274.

[45] X. Zhao, L. Xia, L. Zhang, Z. Ding, D. Yin, J. Tang, Deep reinforcement learning for page-wise recommendations, in: Proceedings of the 12th ACM Conference on Recommender Systems, 2018, pp. 95–103.

[46] O. Gottesman, F. Johansson, M. Komorowski, A. Faisal, D. Sontag, F. Doshi-Velez, et al., Guidelines for reinforcement learning in healthcare, Nature Medicine 25 (1) (2019) 16–18.

[47] M. Minsky, Heuristic Aspects of the Artificial Intelligence Problem, Ed. Services Technical Information agency, 1956.

[48] A.M. Turing, Computing machinery and intelligence, in: Parsing the Turing Test, Springer, 2009, pp. 23–65.

[49] H.K. Ho, L. Zhang, K. Ramamohanarao, S. Martin, A survey of machine learning methods for secondary and supersecondary protein structure prediction, in: Protein Supersecondary Structures, Springer, 2012, pp. 87–106.
[50] P.H. Guzzi, G. Agapito, M. Cannataro, coreSNP: Parallel processing of microarray data, IEEE Transactions on Computers 63 (12) (2014) 2961–2974.
[51] M. Jha, P.H. Guzzi, S. Roy, Qualitative assessment of functional module detectors on microarray and RNASeq data, Network Modeling Analysis in Health Informatics and Bioinformatics 8 (1) (2019) 1–22.
[52] H. Park, K. Maruhashi, R. Yamaguchi, S. Imoto, S. Miyano, Global gene network exploration based on explainable artificial intelligence approach, PLoS ONE 15 (11) (2020) e0241508.
[53] V. Kaul, S. Enslin, S.A. Gross, History of artificial intelligence in medicine, Gastrointestinal Endoscopy 92 (4) (2020) 807–812.
[54] M.E. Moran, Evolution of robotic arms, Journal of Robotic Surgery 1 (2) (2007) 103–111.
[55] B. Kuipers, E.A. Feigenbaum, P.E. Hart, N.J. Nilsson, Shakey: from conception to history, AI Magazine 38 (1) (2017) 88–103.
[56] J. Weizenbaum, Eliza—a computer program for the study of natural language communication between man and machine, Communications of the ACM 9 (1) (1966) 36–45.
[57] S. Weiss, C.A. Kulikowski, A. Safir, Glaucoma consultation by computer, Computers in Biology and Medicine 8 (1) (1978) 25–40.
[58] D. Ferrucci, E. Brown, J. Chu-Carroll, J. Fan, D. Gondek, A.A. Kalyanpur, et al., Building Watson: An overview of the DeepQA project, AI Magazine 31 (3) (2010) 59–79.
[59] N. Bakkar, T. Kovalik, I. Lorenzini, S. Spangler, A. Lacoste, K. Sponaugle, et al., Artificial intelligence in neurodegenerative disease research: use of IBM Watson to identify additional RNA-binding proteins altered in amyotrophic lateral sclerosis, Acta Neuropathologica 135 (2) (2018) 227–247.
[60] C.W. Park, S.W. Seo, N. Kang, B. Ko, B.W. Choi, C.M. Park, et al., Artificial intelligence in health care: Current applications and issues, Journal of Korean Medical Science 35 (42) (2020).
[61] E.J. Lee, Y.H. Kim, N. Kim, D.W. Kang, Deep into the brain: artificial intelligence in stroke imaging, Journal of Stroke 19 (3) (2017) 277.
[62] K. Muhammad, S. Khan, J. Del Ser, V.H.C. de Albuquerque, Deep learning for multi-grade brain tumor classification in smart healthcare systems: A prospective survey, IEEE Transactions on Neural Networks and Learning Systems 32 (2) (2020) 507–522.
[63] V. Mayer-Schönberger, K. Cukier, Big Data: A Revolution That Will Transform How We Live, Work, and Think, Houghton Mifflin Harcourt, 2013.
[64] U.A. Mejias, N. Couldry, Datafication, Internet Policy Review 8 (4) (2019).
[65] M. Ruckenstein, N.D. Schüll, The datafication of health, Annual Review of Anthropology 46 (2017) 261–278.
[66] S.S. Skiena, The Data Science Design Manual, Springer, 2017.
[67] J.D. Kelleher, B. Tierney, Data Science, MIT Press, 2018.
[68] C.J. Wu, Statistics = data science?, Inaugural lecture for the H.C. Carver Chair in Statistics at the University of Michigan, http://www2.isye.gatech.edu/jeffwpresentations/datascience.pdf, 1997.
[69] E.F. Codd, Relational database: A practical foundation for productivity, in: Readings in Artificial Intelligence and Databases, Elsevier, 1989, pp. 60–68.
[70] W.S. McCulloch, W. Pitts, A logical calculus of the ideas immanent in nervous activity, The Bulletin of Mathematical Biophysics 5 (4) (1943) 115–133.

[71] L. Breiman, Statistical modeling: The two cultures (with comments and a rejoinder by the author), Statistical Science 16 (3) (2001) 199–231.
[72] U. Fayyad, G. Piatetsky-Shapiro, P. Smyth, From data mining to knowledge discovery in databases, AI Magazine 17 (3) (1996) 37.
[73] W. Hersh, Information Retrieval: A Biomedical and Health Perspective, Springer, 2020.
[74] A.L. Samuel, Some studies in machine learning using the game of checkers, IBM Journal of Research and Development 3 (3) (1959) 210–229.
[75] J.R. Koza, F.H. Bennett, D. Andre, M.A. Keane, Automated design of both the topology and sizing of analog electrical circuits using genetic programming, in: Artificial Intelligence in Design'96, Springer, 1996, pp. 151–170.
[76] I.H. Witten, E. Frank, M.A. Hall, C.J. Pal, Data Mining: Practical Machine Learning Tools and Techniques, Morgan Kaufmann, 2016.
[77] R. Kitchin, The Data Revolution: Big Data, Open Data, Data Infrastructures and Their Consequences, Sage, 2014.
[78] D.H. Wolpert, The lack of a priori distinctions between learning algorithms, Neural Computation 8 (7) (1996) 1341–1390.
[79] D.H. Wolpert, W.G. Macready, No free lunch theorems for optimization, IEEE Transactions on Evolutionary Computation 1 (1) (1997) 67–82.
[80] Z. Voulgaris, Julia for Data Science, Technics Publications, 2016.
[81] K. Gao, G. Mei, F. Piccialli, S. Cuomo, J. Tu, Z. Huo, Julia language in machine learning: Algorithms, applications, and open issues, Computer Science Review 37 (2020) 100254.
[82] P. Lafaye de Micheaux, R. Drouilhet, B. Liquet, The R Software. Fundamentals of Programming and Statistical Analysis, ISBN 978-1-4614-9019-7, 2013.
[83] J.M. Zelle, Python Programming: An Introduction to Computer Science, Franklin, Beedle & Associates, Inc., 2004.
[84] M. Hall, E. Frank, G. Holmes, B. Pfahringer, P. Reutemann, I.H. Witten, The WEKA data mining software: an update, ACM SIGKDD Explorations Newsletter 11 (1) (2009) 10–18.
[85] M.R. Berthold, N. Cebron, F. Dill, T.R. Gabriel, T. Kötter, T. Meinl, et al., KNIME–the Konstanz information miner: version 2.0 and beyond, ACM SIGKDD Explorations Newsletter 11 (1) (2009) 26–31.
[86] M. Ali, PyCaret: An open source, low-code machine learning library in Python, PyCaret version 1.0.0, https://www.pycaret.org, 2020.
[87] I. Goodfellow, Y. Bengio, A. Courville, Deep Learning, MIT Press, 2016.
[88] D.O. Hebb, The Organization of Behavior: A Neuropsychological Theory, Wiley, 1949.
[89] F. Rosenblatt, The perceptron: a probabilistic model for information storage and organization in the brain, Psychological Review 65 (6) (1958) 386.
[90] B. Widrow, M. Hoff, Adaptive switching circuits, Technical report no. 1553-1, 1960.
[91] D.H. Hubel, T.N. Wiesel, Receptive fields, binocular interaction and functional architecture in the cat's visual cortex, The Journal of Physiology 160 (1) (1962) 106.
[92] K. Fukushima, S. Miyake, T. Ito, Neocognitron: A neural network model for a mechanism of visual pattern recognition, IEEE Transactions on Systems, Man and Cybernetics (5) (1983) 826–834.
[93] M. Minsky, S. Papert, Perceptrons, 1969.
[94] Y. LeCun, et al., Generalization and network design strategies, Connectionism in Perspective 19 (143–155) (1989) 18.

[95] S. Hochreiter, Untersuchungen zu dynamischen neuronalen Netzen, Diploma, Technische Universität München 91 (1) (1991), https://people.idsia.ch/~juergen/SeppHochreiter1991ThesisAdvisorSchmidhuber.pdf.
[96] Y. LeCun, L. Bottou, Y. Bengio, P. Haffner, Gradient-based learning applied to document recognition, Proceedings of the IEEE 86 (11) (1998) 2278–2324.
[97] S. Hochreiter, J. Schmidhuber, Long short-term memory, Neural Computation 9 (8) (1997) 1735–1780.
[98] P. Baldi, Deep learning in biomedical data science, Annual Review of Biomedical Data Science 1 (2018) 181–205.
[99] C. Angermueller, T. Pärnamaa, L. Parts, O. Stegle, Deep learning for computational biology, Molecular Systems Biology 12 (7) (2016) 878.
[100] B. Tang, Z. Pan, K. Yin, A. Khateeb, Recent advances of deep learning in bioinformatics and computational biology, Frontiers in Genetics 10 (2019) 214.
[101] P. Baldi, Deep Learning in Science, Cambridge University Press, 2021.
[102] J.D. Kelleher, Deep Learning, MIT Press, 2019.
[103] T. Hofmann, B. Schölkopf, A.J. Smola, Kernel methods in machine learning, The Annals of Statistics 36 (3) (2008) 1171–1220.
[104] K. Hornik, Approximation capabilities of multilayer feedforward networks, Neural Networks 4 (2) (1991) 251–257.
[105] F. Emmert-Streib, Z. Yang, H. Feng, S. Tripathi, M. Dehmer, An introductory review of deep learning for prediction models with big data, Frontiers in Artificial Intelligence 3 (2020) 4.
[106] Z. Lu, H. Pu, F. Wang, Z. Hu, L. Wang, The expressive power of neural networks: A view from the width, Advances in Neural Information Processing Systems 30 (2017).
[107] J. Angwin, J. Larson, S. Mattu, L. Kirchner, Machine bias: There's software used across the country to predict future criminals. And it's biased against blacks, ProPublica (2016).
[108] R. Caruana, Y. Lou, J. Gehrke, P. Koch, M. Sturm, N. Elhadad, Intelligible models for healthcare: Predicting pneumonia risk and hospital 30-day readmission, in: Proceedings of the 21st ACM SIGKDD International Conference on Knowledge Discovery and Data Mining, ACM, 2015, pp. 1721–1730.
[109] M.T. Ribeiro, S. Singh, C. Guestrin, "Why should I trust you?": Explaining the predictions of any classifier, in: Proceedings of the 22nd ACM SIGKDD International Conference on Knowledge Discovery and Data Mining, ACM, 2016, pp. 1135–1144.
[110] M.T. Ribeiro, S. Singh, C. Guestrin, Model-agnostic interpretability of machine learning, arXiv preprint, arXiv:1606.05386, 2016.
[111] S. Bach, A. Binder, G. Montavon, F. Klauschen, K.R. Müller, W. Samek, On pixel-wise explanations for non-linear classifier decisions by layer-wise relevance propagation, PLoS ONE 10 (7) (2015) e0130140.
[112] A. Shrikumar, P. Greenside, A. Kundaje, Learning important features through propagating activation differences, in: Proceedings of the 34th International Conference on Machine Learning-Volume 70, JMLR.org, 2017, pp. 3145–3153.
[113] M. Al-Shedivat, A. Dubey, E.P. Xing, Contextual explanation networks, arXiv preprint, arXiv:1705.10301, 2017.
[114] G. Montavon, W. Samek, K.R. Müller, Methods for interpreting and understanding deep neural networks, Digital Signal Processing: A Review Journal 73 (2018) 1–15, https://doi.org/10.1016/j.dsp.2017.10.011.
[115] F. Doshi-Velez, B. Kim, Towards a rigorous science of interpretable machine learning, arXiv preprint, arXiv:1702.08608, 2017.

[116] R. Guidotti, A. Monreale, S. Ruggieri, F. Turini, F. Giannotti, D. Pedreschi, A survey of methods for explaining black box models, ACM Computing Surveys (CSUR) 51 (5) (2018) 93.

[117] T. Miller, Explanation in artificial intelligence: Insights from the social sciences, Artificial Intelligence (2018).

[118] G. Ras, P. Haselager, M. van Gerven, Explanation methods in deep learning: Users, values, concerns and challenges, arXiv preprint, arXiv:1803.07517, 2018.

[119] W. Samek, A. Binder, G. Montavon, S. Lapuschkin, K.R. Müller, Evaluating the visualization of what a deep neural network has learned, IEEE Transactions on Neural Networks and Learning Systems 28 (11) (2017) 2660–2673.

[120] L. Arras, G. Montavon, K.R. Müller, W. Samek, Explaining recurrent neural network predictions in sentiment analysis, arXiv preprint, arXiv:1706.07206, 2017.

[121] L.A. Hendricks, Z. Akata, M. Rohrbach, J. Donahue, B. Schiele, T. Darrell, Generating visual explanations, in: European Conference on Computer Vision, Springer, 2016, pp. 3–19.

[122] M. Narayanan, E. Chen, J. He, B. Kim, S. Gershman, F. Doshi-Velez, How do humans understand explanations from machine learning systems? An evaluation of the human-interpretability of explanation, arXiv preprint, arXiv:1802.00682, 2018.

[123] J. van der Waa, J. van Diggelen, K. van den Bosch, M. Neerincx, Contrastive explanations for reinforcement learning in terms of expected consequences, arXiv preprint, arXiv:1807.08706, 2018.

[124] T. Hailesilassie, Rule extraction algorithm for deep neural networks: A review, arXiv preprint, arXiv:1610.05267, 2016.

[125] B. Letham, C. Rudin, T.H. McCormick, D. Madigan, et al., Interpretable classifiers using rules and Bayesian analysis: Building a better stroke prediction model, Annals of Applied Statistics 9 (3) (2015) 1350–1371.

[126] H. Lakkaraju, S.H. Bach, J. Leskovec, Interpretable decision sets: A joint framework for description and prediction, in: Proceedings of the 22nd ACM SIGKDD International Conference on Knowledge Discovery and Data Mining, ACM, 2016, pp. 1675–1684.

[127] J. Adebayo, J. Gilmer, I. Goodfellow, B. Kim, Local explanation methods for deep neural networks lack sensitivity to parameter values, arXiv preprint, arXiv:1810.03307, 2018.

[128] M. Sundararajan, A. Taly, Q. Yan, Axiomatic attribution for deep networks, in: Proceedings of the 34th International Conference on Machine Learning-Volume 70, JMLR.org, 2017, pp. 3319–3328.

[129] A. Shahroudnejad, A. Mohammadi, K.N. Plataniotis, Improved explainability of capsule networks: Relevance path by agreement, arXiv preprint, arXiv:1802.10204, 2018.

[130] D. Doran, S. Schulz, T. Besold, What does explainable AI really mean. A new conceptualization of perspectives, arXiv preprint, arXiv:1710.00794, 2017.

[131] J. Donahue, L. Anne Hendricks, S. Guadarrama, M. Rohrbach, S. Venugopalan, K. Saenko, et al., Long-term recurrent convolutional networks for visual recognition and description, in: Proceedings of the IEEE Conference on Computer Vision and Pattern Recognition, 2015, pp. 2625–2634.

[132] B. Hayes, J.A. Shah, Improving robot controller transparency through autonomous policy explanation, in: Proceedings of the 2017 ACM/IEEE International Conference on Human-Robot Interaction, ACM, 2017, pp. 303–312.

[133] A.A. Sherstov, P. Stone, Improving action selection in MDP's via knowledge transfer, in: AAAI, vol. 5, 2005, pp. 1024–1029.

[134] A. Holzinger, C. Biemann, C.S. Pattichis, D.B. Kell, What do we need to build explainable AI systems for the medical domain?, arXiv preprint, arXiv:1712.09923, 2017.
[135] A. Panchenko, F. Marten, E. Ruppert, S. Faralli, D. Ustalov, S.P. Ponzetto, et al., Unsupervised, knowledge-free, and interpretable word sense disambiguation, arXiv preprint, arXiv:1707.06878, 2017.
[136] J. Clos, N. Wiratunga, S. Massie, Towards explainable text classification by jointly learning lexicon and modifier terms, in: IJCAI-17 Workshop on Explainable AI (XAI), 2017, p. 19.
[137] L. Arras, F. Horn, G. Montavon, K.R. Müller, W. Samek, Explaining predictions of non-linear classifiers in NLP, arXiv preprint, arXiv:1606.07298, 2016.
[138] Y. Zhang, G. Lai, M. Zhang, Y. Zhang, Y. Liu, S. Ma, Explicit factor models for explainable recommendation based on phrase-level sentiment analysis, in: Proceedings of the 37th International ACM SIGIR Conference on Research & Development in Information Retrieval, ACM, 2014, pp. 83–92.
[139] T. Lei, R. Barzilay, T. Jaakkola, Rationalizing neural predictions, arXiv preprint, arXiv: 1606.04155, 2016.
[140] K.M. Tolle, Intelligent Agents, Springer US, Boston, MA, ISBN 978-1-4615-5915-3, 1997, https://doi.org/10.1007/978-1-4615-5915-3_23.
[141] G. Fedele, L. D'Alfonso, G. D'Aquila, Magnetometer bias finite-time estimation using gyroscope data, IEEE Transactions on Aerospace and Electronic Systems 54 (6) (2018) 2926–2936.
[142] G. Fedele, L. D'Alfonso, A coordinates mixing matrix-based model for swarm formation, International Journal of Control 94 (3) (2021) 711–721.
[143] C.W. Reynolds, Flocks, herds and schools: A distributed behavioral model, in: Proceedings of the 14th Annual Conference on Computer Graphics and Interactive Techniques, 1987, pp. 25–34.
[144] W.M. Shen, Learning from the Environment Based on Percepts and Actions, Ph.D. thesis, USA, 1989, AAI9011860.
[145] M. Bera, Artificial Intelligence in Bioinformatics, International Journal of Innovative Science and Research Technology 2 (2) (2021) 433–436, https://ijisrt.com/assets/upload/files/IJISRT21FEB331.pdf.
[146] R. Agarwal, S. Khaitan, S. Sahu, Intelligent agents, in: Distributed Artificial Intelligence, CRC Press, 2020, pp. 19–46.
[147] G. Agapito, C. Pastrello, P.H. Guzzi, I. Jurisica, M. Cannataro, BioPAX-Parser: parsing and enrichment analysis of BioPAX pathways, Bioinformatics (2020) Btaa529, https://doi.org/10.1093/bioinformatics/btaa529, https://academic.oup.com/bioinformatics/article-pdf/doi/10.1093/bioinformatics/btaa529/33263787/btaa529.pdf.
[148] L. Bortolussi, A. Dovier, F. Fogolari, Agent-based protein structure prediction, Multiagent and Grid Systems 3 (2) (2007) 183–197.
[149] Y. Shoham, Agent-oriented programming, Artificial Intelligence 60 (1) (1993) 51–92, https://doi.org/10.1016/0004-3702(93)90034-9, https://www.sciencedirect.com/science/article/pii/0004370293900349.
[150] F. Bellifemine, A. Poggi, G. Rimassa, JADE–a FIPA-compliant agent framework, in: Proceedings of PAAM, vol. 99, London, 1999, p. 33.
[151] L. Braubach, W. Lamersdorf, A. Pokahr, Jadex: Implementing a BDI-infrastructure for JADE agents, 2003.
[152] L.S. Melo, R.F. Sampaio, R.P.S. Leão, G.C. Barroso, J.R. Bezerra, Python-based multi-agent platform for application on power grids, International Transactions on

Electrical Energy Systems 29 (6) (2019) e12012, https://doi.org/10.1002/2050-7038.12012, E12012 ITEES-18-0867.R2, https://onlinelibrary.wiley.com/doi/abs/10.1002/2050-7038.12012, https://onlinelibrary.wiley.com/doi/pdf/10.1002/2050-7038.12012.

[153] J. Lin, E.J. Keogh, S. Lonardi, B.Y. Chiu, A symbolic representation of time series, with implications for streaming algorithms, in: M.J. Zaki, C.C. Aggarwal (Eds.), Proceedings of the 8th ACM SIGMOD Workshop on Research Issues in Data Mining and Knowledge Discovery, DMKD 2003, San Diego, California, USA, June 13, 2003, ACM, 2003, pp. 2–11, https://doi.org/10.1145/882082.882086.

[154] V. Chandola, A. Banerjee, V. Kumar, Anomaly detection: A survey, ACM Computing Surveys (CSUR) 41 (3) (2009) 1–58.

[155] W.R. Taylor, C.A. Orengo, Protein structure alignment, Journal of Molecular Biology 208 (1) (1989) 1–22.

[156] J. Han, M. Kamber, J. Pei, 10 - Cluster analysis: Basic concepts and methods, in: J. Han, M. Kamber, J. Pei (Eds.), Data Mining, third edition, in: The Morgan Kaufmann Series in Data Management Systems, Morgan Kaufmann, Boston, ISBN 978-0-12-381479-1, 2012, pp. 443–495, https://doi.org/10.1016/B978-0-12-381479-1.00010-1, https://www.sciencedirect.com/science/article/pii/B9780123814791000101.

[157] F. Cauteruccio, G. Terracina, D. Ursino, Generalizing identity-based string comparison metrics: Framework and techniques, Knowledge-Based Systems 187 (2020) 104820, https://doi.org/10.1016/j.knosys.2019.06.028, https://www.sciencedirect.com/science/article/pii/S0950705119302953.

[158] F. Cauteruccio, C. Stamile, G. Terracina, D. Ursino, D. Sappey-Marinier, An automated string-based approach to extracting and characterizing white matter fiber-bundles, Computers in Biology and Medicine 77 (2016) 64–75, https://doi.org/10.1016/j.compbiomed.2016.07.015, https://www.sciencedirect.com/science/article/pii/S0010482516301913.

[159] F. Cauteruccio, G. Fortino, A. Guerrieri, G. Terracina, Discovery of hidden correlations between heterogeneous wireless sensor data streams, in: Internet and Distributed Computing Systems, Springer International Publishing, Cham, ISBN 978-3-319-11692-1, 2014, pp. 383–395.

[160] T.H. Cormen, C.E. Leiserson, R.L. Rivest, C. Stein, Introduction to Algorithms, MIT Press, 2009.

[161] R.A. Wagner, M.J. Fischer, The string-to-string correction problem, Journal of the ACM 21 (1) (1974) 168–173, https://doi.org/10.1145/321796.321811.

[162] S.B. Needleman, C.D. Wunsch, A general method applicable to the search for similarities in the amino acid sequence of two proteins, Journal of Molecular Biology 48 (3) (1970) 443–453.

[163] D. Gusfield, Algorithms on stings, trees, and sequences: Computer science and computational biology, ACM SIGACT News 28 (4) (1997) 41–60.

[164] F. Cauteruccio, C. Stamile, G. Terracina, D. Ursino, D. Sappey-Mariniery, An automated string-based approach to white matter fiber-bundles clustering, in: 2015 International Joint Conference on Neural Networks (IJCNN), 2015, pp. 1–8, https://doi.org/10.1109/IJCNN.2015.7280545.

[165] N. Metropolis, A.W. Rosenbluth, M.N. Rosenbluth, A.H. Teller, E. Teller, Equation of state calculations by fast computing machines, Journal of Chemical Physics 21 (6) (1953) 1087–1092.

[166] S. Vinga, J. Almeida, Alignment-free sequence comparison—a review, Bioinformatics 19 (4) (2003) 513–523.

[167] L. Wang, T. Jiang, On the complexity of multiple sequence alignment, Journal of Computational Biology 1 (4) (1994) 337–348.
[168] J. Ren, X. Bai, Y.Y. Lu, K. Tang, Y. Wang, G. Reinert, et al., Alignment-free sequence analysis and applications, Annual Review of Biomedical Data Science 1 (2018) 93–114.
[169] H. Wang, Z. Xu, L. Gao, B. Hao, A fungal phylogeny based on 82 complete genomes using the composition vector method, BMC Evolutionary Biology 9 (1) (2009) 1–13.
[170] S.R. Jun, G.E. Sims, G.A. Wu, S.H. Kim, Whole-proteome phylogeny of prokaryotes by feature frequency profiles: An alignment-free method with optimal feature resolution, Proceedings of the National Academy of Sciences 107 (1) (2010) 133–138.
[171] G. Rizk, D. Lavenier, R. Chikhi, DSK: k-mer counting with very low memory usage, Bioinformatics 29 (5) (2013) 652–653, https://doi.org/10.1093/bioinformatics/btt020, https://academic.oup.com/bioinformatics/article-pdf/29/5/652/702231/btt020.pdf.
[172] X. Robert, P. Gouet, Deciphering key features in protein structures with the new ENDscript server, Nucleic Acids Research 42 (W1) (2014) W320–W324.
[173] Q. Jiang, X. Jin, S.J. Lee, S. Yao, Protein secondary structure prediction: A survey of the state of the art, Journal of Molecular Graphics & Modelling 76 (2017) 379–402.
[174] D. Frishman, P. Argos, Knowledge-based protein secondary structure assignment, Proteins: Structure, Function, and Bioinformatics 23 (4) (1995) 566–579.
[175] W. Kabsch, C. Sander, Dictionary of protein secondary structure: pattern recognition of hydrogen-bonded and geometrical features, Biopolymers: Original Research on Biomolecules 22 (12) (1983) 2577–2637.
[176] J. Garnier, J. Gibrat, B. Robson, GOR secondary structure prediction method version iv, Methods in Enzymology 266 (1996) 540–553.
[177] W. Wardah, M.G. Khan, A. Sharma, M.A. Rashid, Protein secondary structure prediction using neural networks and deep learning: A review, Computational Biology and Chemistry 81 (2019) 1–8.
[178] N. Qian, T.J. Sejnowski, Predicting the secondary structure of globular proteins using neural network models, Journal of Molecular Biology 202 (4) (1988) 865–884.
[179] B. Rost, C. Sander, R. Schneider, PHD–an automatic mail server for protein secondary structure prediction, Computer Applications in the Biosciences 10 (1) (1994) 53–60, http://www.ncbi.nlm.nih.gov/entrez/query.fcgi?cmd=Retrieve&db=pubmed&dopt=Abstract&list_uids=8193956.
[180] J.A. Cuff, M.E. Clamp, A.S. Siddiqui, M. Finlay, G.J. Barton, JPred: a consensus secondary structure prediction server, Bioinformatics 14 (10) (1998) 892–893, https://doi.org/10.1093/bioinformatics/14.10.892, https://academic.oup.com/bioinformatics/article-pdf/14/10/892/9731860/140892.pdf.
[181] C. Mirabello, G. Pollastri, Porter, PaleAle 4.0: high-accuracy prediction of protein secondary structure and relative solvent accessibility, Bioinformatics 29 (16) (2013) 2056–2058.
[182] P. Kukic, C. Mirabello, G. Tradigo, I. Walsh, P. Veltri, G. Pollastri, Toward an accurate prediction of inter-residue distances in proteins using 2D recursive neural networks, BMC Bioinformatics 15 (1) (2014) 1–15.
[183] R. Heffernan, Y. Yang, K. Paliwal, Y. Zhou, Capturing non-local interactions by long short-term memory bidirectional recurrent neural networks for improving prediction of protein secondary structure, backbone angles, contact numbers and solvent accessibility, Bioinformatics 33 (18) (2017) 2842–2849.
[184] F. Pearl, A. Todd, I. Sillitoe, M. Dibley, O. Redfern, T. Lewis, et al., The CATH domain structure database and related resources Gene3D and DHS provide comprehensive do-

main family information for genome analysis, Nucleic Acids Research 33 (Supplement 1) (2005) 247–251, https://doi.org/10.1093/nar/gki024.
[185] A. Murzin, S. Brenner, T. Hubbard, C. Chotia, SCOP: A structural classification of proteins database for the investigation of sequences and structures, Journal of Molecular Biology 247 (1995) 536–540.
[186] C. Orengo, W. Taylor, SSAP: Sequential structure alignment program for protein structure comparisons, Methods in Enzymology 266 (1996) 617–634.
[187] M. Shatsky, Z. Fligelman, R. Nussinov, H. Wolfson, Alignment of flexible protein structures, in: Proc of the 8th International Conference Intelligent Systems for Molecular Biology, 2000, pp. 329–343.
[188] R. Nussinov, H. Wolfson, Efficient detection of three-dimensional structural motifs in biological macromolecules by computer vision techniques, Proceedings of the National Academy of Sciences of the United States of America 88 (1991) 10495–10499.
[189] L. Holm, C. Sander, Searching protein structure databases has come of age, Proteins 19 (165–173) (1994) 165–173.
[190] M. AlQuraishi, AlphaFold at CASP13, Bioinformatics 35 (22) (2019) 4862–4865.
[191] M.L. Metzker, Sequencing technologies – the next generation, Nature Reviews. Genetics 11 (1) (2010) 31–46.
[192] G. Schochetman, C.Y. Ou, W.K. Jones, Polymerase chain reaction, The Journal of Infectious Diseases 158 (6) (1988) 1154–1157.
[193] F. Sanger, S. Nicklen, A.R. Coulson, DNA sequencing with chain-terminating inhibitors, Proceedings of the National Academy of Sciences 74 (12) (1977) 5463–5467.
[194] R. Lowe, N. Shirley, M. Bleackley, S. Dolan, T. Shafee, Transcriptomics technologies, PLoS Computational Biology 13 (5) (2017) e1005457.
[195] Z. Wang, M. Gerstein, M. Snyder, RNA-Seq: a revolutionary tool for transcriptomics, Nature Reviews. Genetics 10 (1) (2009) 57–63.
[196] J. Brown, M. Pirrung, L.A. McCue, FQC dashboard: integrates FastQC results into a web-based, interactive, and extensible FASTQ quality control tool, Bioinformatics 33 (19) (2017) 3137–3139.
[197] N. Joshi, J. Fass, Sickle: A sliding-window, adaptive, quality-based trimming tool for FASTQ files, 2011.
[198] M. Martin, CUTADAPT removes adapter sequences from high-throughput sequencing reads, EMBnet Journal 17 (1) (2011) 10–12.
[199] A. Dobin, C.A. Davis, F. Schlesinger, J. Drenkow, C. Zaleski, S. Jha, et al., STAR: ultrafast universal RNA-seq aligner, Bioinformatics 29 (1) (2013) 15–21.
[200] M. Pertea, G.M. Pertea, C.M. Antonescu, T.C. Chang, J.T. Mendell, S.L. Salzberg, StringTie enables improved reconstruction of a transcriptome from RNA-seq reads, Nature Biotechnology 33 (2015) 290–295.
[201] C. Trapnell, L. Pachter, S.L. Salzberg, TopHat: discovering splice junctions with RNA-seq, Bioinformatics 25 (9) (2009) 1105–1111.
[202] D. Kim, G. Pertea, C. Trapnell, H. Pimentel, R. Kelley, S.L. Salzberg, TopHat2: accurate alignment of transcriptomes in the presence of insertions, deletions and gene fusions, Genome Biology 14 (4) (2013) 1–13.
[203] D. Kim, B. Langmead, S.L. Salzberg, HISAT: a fast spliced aligner with low memory requirements, Nature Methods 12 (4) (2015) 357–360.
[204] S. Anders, P. Pyl, W. Huber, HTSeq—a Python framework to work with high-throughput sequencing data, Bioinformatics 31 (2) (2015) 166–169, https://doi.org/10.1093/bioinformatics/btu638.

[205] R. Patro, S.M. Mount, C. Kingsford, Sailfish enables alignment-free isoform quantification from RNA-seq reads using lightweight algorithms, Nature Biotechnology 32 (5) (2014) 462–464.

[206] N.L. Bray, H. Pimentel, P. Melsted, L. Pachter, Near-optimal probabilistic RNA-seq quantification, Nature Biotechnology 34 (5) (2016) 525–527.

[207] R. Patro, G. Duggal, M.I. Love, R.A. Irizarry, C. Kingsford, Salmon provides fast and bias-aware quantification of transcript expression, Nature Methods 14 (4) (2017) 417–419.

[208] M. Burrows, D. Wheeler, A block-sorting lossless data compression algorithm, in: Digital SRC Research Report, Citeseer, 1994.

[209] M.I. Love, W. Huber, S. Anders, Moderated estimation of fold change and dispersion for RNA-seq data with DESeq2, Genome Biology 15 (2014) 550, https://doi.org/10.1186/s13059-014-0550-8.

[210] P.A. Callinan, A.P. Feinberg, The emerging science of epigenomics, Human Molecular Genetics 15 (suppl_1) (2006) R95–R101.

[211] G. Agapito, P.H. Guzzi, M. Cannataro, A parallel software pipeline for DMET microarray genotyping data analysis, High-Throughput 7 (2) (2018) 17, https://doi.org/10.3390/ht7020017.

[212] P.H. Guzzi, G. Agapito, M.T. Di Martino, M. Arbitrio, P. Tassone, P. Tagliaferri, M. Cannataro, DMET-Miner: Efficient learning of association rules from genotyping data for personalized medicine, in: 2014 IEEE International Conference on Bioinformatics and Biomedicine (BIBM), IEEE, 2014, pp. 59–62, https://doi.org/10.1109/BIBM.2014.6999127.

[213] G. Agapito, M. Cannataro, P.H. Guzzi, F. Marozzo, D. Talia, P. Trunfio, Cloud4SNP: Distributed analysis of SNP microarray data on the cloud, in: Proceedings of the International Conference on Bioinformatics, Computational Biology and Biomedical Informatics, BCB'13, Association for Computing Machinery, New York, NY, USA, ISBN 9781450324342, 2013, pp. 468–475, https://doi.org/10.1145/2506583.2506605.

[214] G. Agapito, P.H. Guzzi, M. Cannataro, An efficient and scalable SPARK preprocessing methodology for Genome Wide Association Studies, in: 28th Euromicro International Conference on Parallel, Distributed and Network-Based Processing (PDP), IEEE, 2020, pp. 369–375, https://doi.org/10.1109/PDP50117.2020.00063.

[215] M.S. Kim, S.M. Pinto, D. Getnet, R.S. Nirujogi, S.S. Manda, R. Chaerkady, et al., A draft map of the human proteome, Nature 509 (7502) (2014) 575–581.

[216] J.M. Walker, The Proteomics Protocols Handbook, Springer, 2005.

[217] A. Brückner, C. Polge, N. Lentze, D. Auerbach, U. Schlattner, Yeast two-hybrid, a powerful tool for systems biology, International Journal of Molecular Sciences 10 (6) (2009) 2763–2788.

[218] Y.V. Miteva, H.G. Budayeva, I.M. Cristea, Proteomics-based methods for discovery, quantification, and validation of protein–protein interactions, Analytical Chemistry 85 (2) (2013) 749–768.

[219] G.T. Hart, I. Lee, E.M. Marcotte, A high-accuracy consensus map of yeast protein complexes reveals modular nature of gene essentiality, BMC Bioinformatics 8 (1) (2007) 1–11.

[220] M. Cannataro, P.H. Guzzi, A. Sarica, Data mining and life sciences applications on the grid, Wiley Interdisciplinary Reviews: Data Mining and Knowledge Discovery 3 (3) (2013) 216–238, https://doi.org/10.1002/widm.1090.

[221] M. Cannataro, P.H. Guzzi, P. Veltri, Protein-to-protein interactions: Technologies, databases, and algorithms, ACM Computing Surveys 43 (2010) 1–36, https://doi.org/10.1145/1824795.1824796.

[222] Y. Moreau, L.C. Tranchevent, Computational tools for prioritizing candidate genes: boosting disease gene discovery, Nature Reviews. Genetics 13 (8) (2012) 523–536.

[223] M.A. Harris, J. Clark, A. Ireland, J. Lomax, M. Ashburner, et al., The Gene Ontology (GO) database and informatics resource, Nucleic Acids Research 32 (Database issue) (2004) 258–261.

[224] P.N. Robinson, S. Kohler, S. Bauer, D. Seelow, D. Horn, S. Mundlos, The human phenotype ontology: a tool for annotating and analyzing human hereditary disease, American Journal of Human Genetics 83 (5) (2008) 610–615.

[225] L.M. Schriml, C. Arze, S. Nadendla, Y.W.W. Chang, M. Mazaitis, V. Felix, et al., Disease ontology: a backbone for disease semantic integration, Nucleic Acids Research 40 (D1) (2012) D940–D946.

[226] L.C. Tranchevent, F.B. Capdevila, D. Nitsch, B. De Moor, P. De Causmaecker, Y. Moreau, et al., A guide to web tools to prioritize candidate genes, Briefings in Bioinformatics 12 (1) (2010) 22–32, https://doi.org/10.1093/bib/bbq007.

[227] A.M. Liekens, J. De Knijf, W. Daelemans, B. Goethals, P. De Rijk, J. Del-Favero, Biograph: unsupervised biomedical knowledge discovery via automated hypothesis generation, Genome Biology 12 (6) (2011) 1–12.

[228] L.C. Tranchevent, A. Ardeshirdavani, S. ElShal, D. Alcaide, J. Aerts, D. Auboeuf, et al., Candidate gene prioritization with endeavour, Nucleic Acids Research 44 (W1) (2016) W117–W121.

[229] A. Schlicker, T. Lengauer, M. Albrecht, Improving disease gene prioritization using the semantic similarity of gene ontology terms, Bioinformatics 26 (18) (2010) i561–i567, https://doi.org/10.1093/bioinformatics/btq384.

[230] J. Chen, E.E. Bardes, B.J. Aronow, A.G. Jegga, ToppGene suite for gene list enrichment analysis and candidate gene prioritization, Nucleic Acids Research 37 (suppl 2) (2009) W305–W311, https://doi.org/10.1093/nar/gkp427.

[231] M. Cannataro, P.H. Guzzi, M. Milano, GoD: An R-package based on ontologies for prioritization of genes with respect to diseases, Journal of Computational Science 9 (2015) 7–13.

[232] Gene Ontology Consortium, The Gene Ontology resource: enriching a GOld mine, Nucleic Acids Research 49 (D1) (2021) D325–D334, https://doi.org/10.1093/nar/gkaa1113.

[233] S. Köhler, M. Gargano, N. Matentzoglu, L.C. Carmody, D. Lewis-Smith, N.A. Vasilevsky, et al., The Human Phenotype Ontology in 2021, Nucleic Acids Research 49 (D1) (2021) D1207–D1217, https://doi.org/10.1093/nar/gkaa1043.

[234] L.M. Schriml, E. Mitraka, J. Munro, B. Tauber, M. Schor, L. Nickle, et al., Human Disease Ontology 2018 update: classification, content and workflow expansion, Nucleic Acids Research 47 (D1) (2018) D955–D962, https://doi.org/10.1093/nar/gky1032.

[235] C. Pesquita, D. Faria, A.O. Falcao, P. Lord, F.M. Couto, Semantic similarity in biomedical ontologies, PLoS Computational Biology 5 (7) (2009) e1000443, https://doi.org/10.1371/journal.pcbi.1000443.

[236] H. Wang, H. Zheng, F. Azuaje, Ontology- and graph-based similarity assessment in biological networks, Bioinformatics 26 (20) (2010) 2643–2644, https://doi.org/10.1093/bioinformatics/btq477.

[237] L. du Plessis, N. Skunca, C. Dessimoz, The what, where, how and why of gene ontology–a primer for bioinformaticians, Briefings in Bioinformatics 12 (6) (2011) 723–735, https://doi.org/10.1093/bib/bbr002.

[238] P.H. Guzzi, M. Mina, C. Guerra, M. Cannataro, Semantic similarity analysis of protein data: assessment with biological features and issues, Briefings in Bioinformatics 13 (5) (2012) 569–585.

[239] P. Resnik, Using information content to evaluate semantic similarity in a taxonomy, in: IJCAI, 1995, pp. 448–453, http://citeseerx.ist.psu.edu/viewdoc/summary?doi=10.1.1.55.5277.

[240] D. Lin, An information-theoretic definition of similarity, in: Proceedings of the 15th International Conference on Machine Learning, Morgan Kaufmann, San Francisco, CA, 1998, http://citeseerx.ist.psu.edu/viewdoc/summary?doi=10.1.1.55.1832.

[241] C. Pesquita, D. Faria, H. Bastos, A. Ferreira, A. Falcao, F. Couto, Metrics for GO based protein semantic similarity: a systematic evaluation, BMC Bioinformatics 9 (Suppl 5) (2008) S4+, https://doi.org/10.1186/1471-2105-9-s5-s4.

[242] Y.R. Cho, M. Mina, Y. Lu, N. Kwon, P.H. Guzzi, M-Finder: Uncovering functionally associated proteins from interactome data integrated with GO annotations, Proteome Science 11 (Suppl 1) (2013) S3.

[243] D. Faria, A. Schlicker, C. Pesquita, H. Bastos, A.E. Ferreira, M. Albrecht, et al., Mining go annotations for improving annotation consistency, PLoS ONE 7 (7) (2012) e40519.

[244] P. Manda, F. McCarthy, S.M. Bridges, Interestingness measures and strategies for mining multi-ontology multi-level association rules from gene ontology annotations for the discovery of new go relationships, Journal of Biomedical Informatics 46 (5) (2013) 849–856.

[245] G. Agapito, M. Milano, P.H. Guzzi, M. Cannataro, Improving annotation quality in gene ontology by mining cross-ontology weighted association rules, in: Bioinformatics and Biomedicine (BIBM), 2014 IEEE International Conference on, IEEE, 2014, pp. 1–8.

[246] G. Agapito, M. Cannataro, P.H. Guzzi, M. Milano, Using GO-WAR for mining cross-ontology weighted association rules, Computer Methods and Programs in Biomedicine 120 (2) (2015) 113–122.

[247] G. Agapito, M. Milano, P.H. Guzzi, M. Cannataro, Efficient learning of association rules from human phenotype ontology, in: Proceedings of the 6th ACM Conference on Bioinformatics, Computational Biology and Health Informatics, 2015, pp. 568–573.

[248] G. Agapito, M. Milano, P.H. Guzzi, M. Cannataro, Mining association rules from disease ontology, in: 2019 IEEE International Conference on Bioinformatics and Biomedicine (BIBM), IEEE, 2019, pp. 2239–2243.

[249] D.W. Huang, B.T. Sherman, R.A. Lempicki, Bioinformatics enrichment tools: paths toward the comprehensive functional analysis of large gene lists, Nucleic Acids Research 37 (1) (2009) 1–13.

[250] GO Consortium, Gene ontology consortium: going forward, Nucleic Acids Research 43 (D1) (2015) D1049–D1056.

[251] H. Tipney, L. Hunter, An introduction to effective use of enrichment analysis software, Human Genomics 4 (3) (2010) 1–5.

[252] S. Roy, P.H. Guzzi, Biological network inference from microarray data, current solutions, and assessments, in: Microarray Data Analysis, Springer, 2015, pp. 155–167.

[253] M. Wilm, Quantitative proteomics in biological research, Proteomics 9 (20) (2009) 4590–4605, https://doi.org/10.1002/pmic.200900299.

[254] A. Sarica, C. Critelli, P.H. Guzzi, A. Cerasa, A. Quattrone, M. Cannataro, Application of different classification techniques on brain morphological data, in: Proceedings of the 26th IEEE International Symposium on Computer-Based Medical Systems, IEEE, 2013, pp. 425–428, https://doi.org/10.1109/CBMS.2013.6627832.
[255] M. Cannataro, P.H. Guzzi, Data Management of Protein Interaction Networks, Wiley Series in Bioinformatics, Wiley-IEEE Computer Society Press, Hoboken, NJ, ISBN 978-1-118-10374-6, 2011, https://doi.org/10.1002/9781118103746.
[256] M. Mina, P. Guzzi, Improving the robustness of local network alignment: Design and extensive assessment of a Markov clustering-based approach, IEEE/ACM Transactions on Computational Biology and Bioinformatics 11 (3) (2014) 561–572, https://doi.org/10.1109/TCBB.2014.2318707.
[257] A. Schrattenholz, K. Groebe, V. Soskic, Systems biology approaches and tools for analysis of interactomes and multi-target drugs, in: J.M. Walker, Q. Yan (Eds.), Systems Biology in Drug Discovery and Development, in: Methods in Molecular Biology, vol. 662, Humana Press, Totowa, NJ, ISBN 978-1-60761-799-0, 2010, pp. 29–58, https://doi.org/10.1007/978-1-60761-800-3_2, chap. 2.
[258] A.L. Barabasi, Z.N. Oltvai, Network biology: understanding the cell's functional organization, Nature Reviews. Genetics 5 (2) (2004) 101–113, https://doi.org/10.1038/nrg1272.
[259] M. Kang, E. Ko, T.B. Mersha, A roadmap for multi-omics data integration using deep learning, Briefings in Bioinformatics 23 (1) (2021) Bbab454, https://doi.org/10.1093/bib/bbab454, https://academic.oup.com/bib/article-pdf/23/1/bbab454/42231062/bbab454.pdf.
[260] M.E. Gallo Cantafio, K. Grillone, D. Caracciolo, F. Scionti, M. Arbitrio, V. Barbieri, et al., From single level analysis to multi-omics integrative approaches: a powerful strategy towards the precision oncology, High-Throughput 7 (4) (2018) 33.
[261] P.S. Reel, S. Reel, E. Pearson, E. Trucco, E. Jefferson, Using machine learning approaches for multi-omics data analysis: A review, Biotechnology Advances (2021) 107739.
[262] M.V. Iorio, C.M. Croce, microRNA involvement in human cancer, Carcinogenesis 33 (6) (2012) 1126–1133, https://doi.org/10.1093/carcin/bgs140, http://carcin.oxfordjournals.org/content/33/6/1126.abstract, http://carcin.oxfordjournals.org/content/33/6/1126.full.pdf+html.
[263] M. Lionetti, P. Musto, M.T. Di Martino, S. Fabris, L. Agnelli, K. Todoerti, et al., Biological and clinical relevance of miRNA expression signatures in primary plasma cell leukemia, Clinical Cancer Research 19 (12) (2013) 3130–3142.
[264] F. Ortuso, D. Mercatelli, P.H. Guzzi, F.M. Giorgi, Structural genetics of circulating variants affecting the SARS-CoV-2 spike/human ACE2 complex, Journal of Biomolecular Structure & Dynamics (2021) 1–11.
[265] Z. Yan, P.K. Shah, S.B. Amin, M.K. Samur, N. Huang, X. Wang, et al., Integrative analysis of gene and miRNA expression profiles with transcription factor–miRNA feed-forward loops identifies regulators in human cancers, Nucleic Acids Research (2012) gks395.
[266] A. Bisognin, G. Sales, A. Coppe, S. Bortoluzzi, C. Romualdi, MAGIA 2: from miRNA and genes expression data integrative analysis to microRNA–transcription factor mixed regulatory circuits (2012 update), Nucleic Acids Research (2012) gks460.
[267] G.T. Huang, C. Athanassiou, P.V. Benos, mirConnX: condition-specific mRNA-microRNA network integrator, Nucleic Acids Research (2011), https://doi.org/10.1093/

nar/gkr276, http://nar.oxfordjournals.org/content/early/2011/05/10/nar.gkr276.abstract, http://nar.oxfordjournals.org/content/early/2011/05/10/nar.gkr276.full.pdf+html.

[268] A.S. Afshar, J. Xu, J. Goutsias, Integrative identification of deregulated miRNA/TF-mediated gene regulatory loops and networks in prostate cancer, PLoS ONE 9 (6) (2014) e100806, https://doi.org/10.1371/journal.pone.0100806.

[269] G.K. Smyth, limma: Linear models for microarray data, in: R. Gentleman, V. Carey, W. Huber, R. Irizarry, S. Dudoit (Eds.), Bioinformatics and Computational Biology Solutions Using R and Bioconductor, in: Statistics for Biology and Health, Springer New York, New York, ISBN 0-387-25146-4, 2005, pp. 397–420, https://doi.org/10.1007/0-387-29362-0_23, chap. 23.

[270] M. Settino, A. Bernasconi, G. Ceddia, G. Agapito, M. Masseroli, M. Cannataro, Using GMQL-Web for querying, downloading and integrating public with private genomic datasets, in: X.M. Shi, M. Buck, J. Ma, P. Veltri (Eds.), Proceedings of the 10th ACM International Conference on Bioinformatics, Computational Biology and Health Informatics, BCB 2019, Niagara Falls, NY, USA, September 7-10, 2019, ACM, 2019, pp. 688–693, https://doi.org/10.1145/3307339.3343466.

[271] L. Kannan, M. Ramos, A. Re, N. El-Hachem, Z. Safikhani, D.M. Gendoo, et al., Public data and open source tools for multi-assay genomic investigation of disease, Briefings in Bioinformatics 17 (4) (2016) 603–615, https://doi.org/10.1093/bib/bbv080.

[272] M. Masseroli, A. Canakoglu, P. Pinoli, A. Kaitoua, A. Gulino, O. Horlova, et al., Processing of big heterogeneous genomic datasets for tertiary analysis of next generation sequencing data, Bioinformatics (2018) bty688, https://doi.org/10.1093/bioinformatics/bty688, https://academic.oup.com/bioinformatics/article/35/5/729/5067860.

[273] K. Tomczak, P. Czerwinska, M. Wiznerowicz, The Cancer Genome Atlas (TCGA): an immeasurable source of knowledge, Contemporary Oncology (Pozn) 19 (1A) (2015) 68–77.

[274] E. Cerami, J. Gao, U. Dogrusoz, B.E. Gross, S.O. Sumer, B.A. Aksoy, et al., The cBio cancer genomics portal: an open platform for exploring multidimensional cancer genomics data, Cancer Discovery 2 (5) (2012) 401–404.

[275] F. Cumbo, G. Fiscon, S. Ceri, M. Masseroli, E. Weitschek, TCGA2BED: extracting, extending, integrating, and querying The Cancer Genome Atlas, BMC Bioinformatics 18 (1) (2017) 6.

[276] M. Masseroli, P. Pinoli, F. Venco, A. Kaitoua, V. Jalili, F. Palluzzi, et al., GenoMetric Query Language: a novel approach to large-scale genomic data management, Bioinformatics 31 (12) (2015) 1881–1888.

[277] M. Settino, M. Cannataro, Survey of main tools for querying and analyzing TCGA data, in: 2018 IEEE International Conference on Bioinformatics and Biomedicine (BIBM), IEEE, 2018, pp. 1711–1718, https://doi.org/10.1109/BIBM.2018.8621270.

[278] M. Settino, M. Arbitrio, F. Scionti, D. Caracciolo, G. Agapito, P. Tassone, et al., Identifying prognostic markers for multiple myeloma through integration and analysis of MMRF-CoMMpass data, Journal of Computational Science 51 (2021) 101346, https://doi.org/10.1016/j.jocs.2021.101346.

[279] P. Guzzi, M. Cannataro, Micro-analyzer: a tool for automatic pre-processing of multiple Affymetrix arrays, EMBnet Journal 18 (2012).

[280] F. Gullo, G. Ponti, A. Tagarelli, G. Tradigo, P. Veltri, A time series approach for clustering mass spectrometry data, Journal of Computational Science 3 (5) (2012) 344–355, https://doi.org/10.1016/j.jocs.2011.06.008, Advanced Computing Solutions for Health Care and Medicine, http://www.sciencedirect.com/science/article/pii/S1877750311000627.

[281] M. Mina, P.H. Guzzi, Improving the robustness of local network alignment: design and extensive assessment of a Markov clustering-based approach, IEEE/ACM Transactions on Computational Biology and Bioinformatics 11 (3) (2014) 561–572.
[282] A. Fornito, A. Zalesky, M. Breakspear, Graph analysis of the human connectome: promise, progress, and pitfalls, NeuroImage 80 (2013) 426–444.
[283] M. Milano, P.H. Guzzi, M. Cannataro, Network building and analysis in connectomics studies: a review of algorithms, databases and technologies, Network Modeling Analysis in Health Informatics and Bioinformatics 8 (2019) 13, https://doi.org/10.1007/s13721-019-0192-6.
[284] T. Nepusz, A. Paccanaro, Structural pattern discovery in protein–protein interaction networks, in: Springer Handbook of Bio-/Neuroinformatics, Springer, 2014, pp. 375–398.
[285] U. Stelzl, U. Worm, M. Lalowski, C. Haenig, F.H. Brembeck, H. Goehler, et al., A human protein-protein interaction network: a resource for annotating the proteome, Cell 122 (6) (2005) 957–968.
[286] O. Sporns, G. Tononi, R. Kötter, The human connectome: a structural description of the human brain, PLoS Computational Biology 4 (4) (2005) e42.
[287] M. Cannataro, P.H. Guzzi, P. Veltri, IMPRECO: Distributed prediction of protein complexes, Future Generation Computer Systems (2009), https://doi.org/10.1016/j.future.2009.08.001.
[288] M. Milano, W. Hayes, P. Veltri, M. Cannataro, P.H. Guzzi, SL-GLAlign: improving local alignment of biological networks through simulated annealing, Network Modeling Analysis in Health Informatics and Bioinformatics 9 (10) (2020) 1–16, https://doi.org/10.1007/s13721-019-0214-4.
[289] E. Bullmore, O. Sporns, Complex brain networks: graph theoretical analysis of structural and functional systems, Nature Reviews. Neuroscience 10 (3) (2009) 186–198.
[290] P.H. Guzzi, T. Milenković, Survey of local and global biological network alignment: the need to reconcile the two sides of the same coin, Briefings in Bioinformatics (2017) bbw132.
[291] M. Milano, P.H. Guzzi, O. Tymofieva, D. Xu, C. Hess, P. Veltri, et al., An extensive assessment of network alignment algorithms for comparison of brain connectomes, BMC Bioinformatics 18 (6) (2017) 235.
[292] M. Milano, T. Milenković, M. Cannataro, P.H. Guzzi, L-HetNetAligner: a novel algorithm for local alignment of heterogeneous biological networks, Scientific Reports 10 (1) (2020) 1–20.
[293] S. Tavazoie, J.D. Hughes, M.J. Campbell, R.J. Cho, G.M. Church, Systematic determination of genetic network architecture, Nature Genetics 22 (3) (1999) 281–285.
[294] H. Jeong, B. Tombor, R. Albert, Z.N. Oltvai, A.L. Barabási, The large-scale organization of metabolic networks, Nature 407 (6804) (2000) 651–654.
[295] S. Roy, D. Das, D. Choudhury, G.G. Gohain, R. Sharma, D.K. Bhattacharyya, Causality inference techniques for in-silico gene regulatory network, in: Mining Intelligence and Knowledge Exploration, Springer, 2013, pp. 432–443.
[296] J. De Las Rivas, C. Fontanillo, Protein–protein interactions essentials: key concepts to building and analyzing interactome networks, PLoS Computational Biology 6 (6) (2010) e1000807.
[297] P. Erdos, A. Rényi, On the evolution of random graphs, Publications of the Mathematical Institute of the Hungarian Academy of Sciences 5 (1) (1960) 17–60.
[298] D. Fell, A. Wagner, The small world of metabolism, Nature Biotechnology 18 (11) (2000) 1121–1122.

[299] A.L. Barabási, R. Albert, Emergence of scaling in random networks, Science 286 (5439) (1999) 509–512.

[300] M. Penrose, et al., Random Geometric Graphs. No. 5, Oxford University Press, 2003.

[301] M. Penrose, Geometric Random Graphs, Oxford University Press, Oxford, 2003.

[302] M. Cannataro, P.H. Guzzi, T. Mazza, G. Tradigo, P. Veltri, Preprocessing of mass spectrometry proteomics data on the grid, in: 18th IEEE Symposium on Computer-Based Medical Systems (CBMS'05), IEEE, 2005, pp. 549–554.

[303] P.H. Guzzi, M. Cannataro, μ-CS: An extension of the TM4 platform to manage Affymetrix binary data, BMC Bioinformatics 11 (1) (2010) 315.

[304] W.L. Hamilton, R. Ying, J. Leskovec, Representation learning on graphs: Methods and applications, arXiv preprint, arXiv:1709.05584, 2017.

[305] P. Cui, X. Wang, J. Pei, W. Zhu, A survey on network embedding, IEEE Transactions on Knowledge and Data Engineering 31 (5) (2018) 833–852.

[306] C. Su, J. Tong, Y. Zhu, P. Cui, F. Wang, Network embedding in biomedical data science, Briefings in Bioinformatics 21 (1) (2020) 182–197.

[307] W. Nelson, M. Zitnik, B. Wang, J. Leskovec, A. Goldenberg, R. Sharan, To embed or not: network embedding as a paradigm in computational biology, Frontiers in Genetics 10 (2019).

[308] P. Goyal, E. Ferrara, Graph embedding techniques, applications, and performance: A survey, Knowledge-Based Systems 151 (2018) 78–94.

[309] C. Zucco, B. Calabrese, G. Agapito, P.H. Guzzi, M. Cannataro, Sentiment analysis for mining texts and social networks data: Methods and tools, Wiley Interdisciplinary Reviews: Data Mining and Knowledge Discovery 10 (1) (2020) e1333.

[310] S. Cao, W. Lu, Q. Xu, GraRep: Learning graph representations with global structural information, in: Proceedings of the 24th ACM International on Conference on Information and Knowledge Management, 2015, pp. 891–900.

[311] M. Ou, P. Cui, J. Pei, Z. Zhang, W. Zhu, Asymmetric transitivity preserving graph embedding, in: Proceedings of the 22nd ACM SIGKDD International Conference on Knowledge Discovery and Data Mining, 2016, pp. 1105–1114.

[312] B. Perozzi, R. Al-Rfou, S. Skiena, DeepWalk: Online learning of social representations, in: Proceedings of the 20th ACM SIGKDD International Conference on Knowledge Discovery and Data Mining, 2014, pp. 701–710.

[313] A. Grover, J. Leskovec, node2vec: Scalable feature learning for networks, in: Proceedings of the 22nd ACM SIGKDD International Conference on Knowledge Discovery and Data Mining, 2016, pp. 855–864.

[314] J. Tang, M. Qu, M. Wang, M. Zhang, J. Yan, Q. Mei, Line: Large-scale information network embedding, in: Proceedings of the 24th International Conference on World Wide Web, 2015, pp. 1067–1077.

[315] S.F. Altschul, T.L. Madden, A.A. Schäffer, J. Zhang, Z. Zhang, W. Miller, et al., Gapped BLAST and PSI-BLAST: a new generation of protein database search programs, Nucleic Acids Research 25 (17) (1997) 3389–3402.

[316] J. Berg, M. Lässig, Local graph alignment and motif search in biological networks, Proceedings of the National Academy of Sciences of the United States of America 101 (41) (2004) 14689–14694.

[317] G. Ciriello, M. Mina, P.H. Guzzi, M. Cannataro, C. Guerra, AlignNemo: a local network alignment method to integrate homology and topology, PLoS ONE 7 (6) (2012) e38107.

[318] Y. Tian, R.C. Mceachin, C. Santos, D.J. States, J.M. Patel, Saga: a subgraph matching tool for biological graphs, Bioinformatics 23 (2) (2006) 232–239.

[319] L.A. Zager, G.C. Verghese, Graph similarity scoring and matching, Applied Mathematics Letters 21 (1) (2008) 86–94.
[320] R. Raveaux, J.C. Burie, J.M. Ogier, A graph matching method and a graph matching distance based on subgraph assignments, Pattern Recognition Letters 31 (5) (2010) 394–406.
[321] J. Kobler, U. Schöning, J. Torán, The Graph Isomorphism Problem: Its Structural Complexity, Springer Science & Business Media, 2012.
[322] V. Saraph, T. Milenković, MAGNA: maximizing accuracy in global network alignment, Bioinformatics 30 (20) (2014) 2931–2940.
[323] V. Vijayan, V. Saraph, T. Milenković, MAGNA++: Maximizing accuracy in global network alignment via both node and edge conservation, Bioinformatics 31 (14) (2015) 2409–2411.
[324] N. Mamano, W. Hayes, Sana: Simulated annealing network alignment applied to biological networks, arXiv preprint, arXiv:1607.02642, 2016.
[325] N. Malod-Dognin, K. Ban, N. Pržulj, Unified alignment of protein-protein interaction networks, Scientific Reports 7 (1) (2017) 953.
[326] C.S. Liao, K. Lu, M. Baym, R. Singh, B. Berger, IsoRankN: spectral methods for global alignment of multiple protein networks, Bioinformatics 25 (12) (2009) i253–i258.
[327] O. Kuchaiev, T. Milenković, V. Memišević, W. Hayes, N. Pržulj, Topological network alignment uncovers biological function and phylogeny, Journal of the Royal Society Interface 7 (50) (2010) 1341–1354.
[328] T. Milenković, W.L. Ng, W. Hayes, N. Pržulj, Optimal network alignment with graphlet degree vectors, Cancer Informatics 9 (2010), CIN–S4744.
[329] O. Kuchaiev, N. Pržulj, Integrative network alignment reveals large regions of global network similarity in yeast and human, Bioinformatics 27 (10) (2011) 1390–1396.
[330] V. Memišević, N. Pržulj, C-GRAAL: common-neighbors-based global GRAph ALignment of biological networks, Integrative Biology 4 (7) (2012) 734–743.
[331] N. Malod-Dognin, N. Pržulj, L-GRAAL: Lagrangian graphlet-based network aligner, Bioinformatics 31 (13) (2015) 2182–2189.
[332] R. Patro, C. Kingsford, Global network alignment using multiscale spectral signatures, Bioinformatics 28 (23) (2012) 3105–3114.
[333] Y. Sun, J. Crawford, J. Tang, T. Milenković, Simultaneous optimization of both node and edge conservation in network alignment via wave, in: International Workshop on Algorithms in Bioinformatics, Springer, 2015, pp. 16–39.
[334] L. Meng, A. Striegel, T. Milenkovic, IGLOO: Integrating global and local biological network alignment, arXiv preprint, arXiv:1604.06111, 2016.
[335] A.J. Enright, S. Van Dongen, C.A. Ouzounis, An efficient algorithm for large-scale detection of protein families, Nucleic Acids Research 30 (7) (2002) 1575–1584.
[336] M. Milano, P.H. Guzzi, M. Cannataro, GLAlign: A novel algorithm for local network alignment, IEEE/ACM Transactions on Computational Biology and Bioinformatics (2018).
[337] R. Sharan, T. Ideker, Modeling cellular machinery through biological network comparison, Nature Biotechnology 24 (4) (2006) 427–433, https://doi.org/10.1038/nbt1196, http://www.ncbi.nlm.nih.gov/pubmed/16601728.
[338] R.A. Pache, A. Ceol, P. Aloy, NetAligner: a network alignment server to compare complexes, pathways and whole interactomes, Nucleic Acids Research 40 (W1) (2012) W157–W161.
[339] V. Vijayan, T. Milenković, Multiple network alignment via multiMAGNA++, IEEE/ACM Transactions on Computational Biology and Bioinformatics (2017).

[340] R. Ibragimov, M. Malek, J. Baumbach, J. Guo, Multiple graph edit distance: simultaneous topological alignment of multiple protein-protein interaction networks with an evolutionary algorithm, in: Proceedings of the 2014 Annual Conference on Genetic and Evolutionary Computation, ACM, 2014, pp. 277–284.

[341] J. Hu, K. Reinert, LocalAli: an evolutionary-based local alignment approach to identify functionally conserved modules in multiple networks, Bioinformatics 31 (3) (2014) 363–372.

[342] S.M.E. Sahraeian, B.J. Yoon, Smetana: accurate and scalable algorithm for probabilistic alignment of large-scale biological networks, PLoS ONE 8 (7) (2013) e67995.

[343] V. Gligorijević, N. Malod-Dognin, N. Pržulj, Fuse: multiple network alignment via data fusion, Bioinformatics 32 (8) (2016) 1195–1203.

[344] J. Hu, B. Kehr, K. Reinert, NetCoffee: a fast and accurate global alignment approach to identify functionally conserved proteins in multiple networks, Bioinformatics 30 (4) (2014) 540–548.

[345] F. Alkan, C. Erten, BEAMS: backbone extraction and merge strategy for the global many-to-many alignment of multiple PPI networks, Bioinformatics 30 (4) (2014) 531–539.

[346] V. Vijayan, S. Gu, E.T. Krebs, L. Meng, T. Milenkovic, Pairwise versus multiple global network alignment, IEEE Access 8 (2020) 41961–41974.

[347] L. Meng, A. Striegel, T. Milenković, Local versus global biological network alignment, Bioinformatics 32 (20) (2016) 3155–3164.

[348] M. Kanehisa, S. Goto, KEGG: Kyoto encyclopedia of genes and genomes, Nucleic Acids Research 28 (1) (2000) 27–30.

[349] R. Caspi, H. Foerster, C.A. Fulcher, P. Kaipa, M. Krummenacker, M. Latendresse, et al., The MetaCyc database of metabolic pathways and enzymes and the BioCyc collection of pathway/genome databases, Nucleic Acids Research 36 (suppl_1) (2007) D623–D631.

[350] H. Mi, A. Muruganujan, D. Ebert, X. Huang, P.D. Thomas, Panther version 14: more genomes, a new panther go-slim and improvements in enrichment analysis tools, Nucleic Acids Research 47 (D1) (2019) D419–D426.

[351] E.G. Cerami, B.E. Gross, E. Demir, I. Rodchenkov, Ö Babur, N. Anwar, et al., Pathway commons, a web resource for biological pathway data, Nucleic Acids Research 39 (suppl_1) (2010) D685–D690.

[352] A. Fabregat, K. Sidiropoulos, G. Viteri, O. Forner, P. Marin-Garcia, V. Arnau, et al., Reactome pathway analysis: a high-performance in-memory approach, BMC Bioinformatics 18 (1) (2017) 142, https://doi.org/10.1186/s12859-017-1559-2.

[353] A. Fabregat, S. Jupe, L. Matthews, K. Sidiropoulos, M. Gillespie, P. Garapati, et al., The reactome pathway knowledgebase, Nucleic Acids Research 46 (D1) (2017) D649–D655.

[354] A.R. Pico, T. Kelder, M.P. Van Iersel, K. Hanspers, B.R. Conklin, C. Evelo, WikiPathways: pathway editing for the people, PLoS Biology 6 (7) (2008) e184.

[355] E. Demir, M.P. Cary, S. Paley, K. Fukuda, C. Lemer, I. Vastrik, et al., The BioPAX community standard for pathway data sharing, Nature Biotechnology 28 (9) (2010) 935–942.

[356] C.M. Lloyd, M.D. Halstead, P.F. Nielsen, CellML: its future, present and past, Progress in Biophysics and Molecular Biology 85 (2–3) (2004) 433–450.

[357] L. Perfetto, M.L. Acencio, G. Bradley, G. Cesareni, N. Del Toro, D. Fazekas, et al., CausalTAB: the PSI-MITAB 2.8 updated format for signalling data representation and dissemination, Bioinformatics 35 (19) (2019) 3779–3785.

[358] A.C. Zambon, S. Gaj, I. Ho, K. Hanspers, K. Vranizan, C.T. Evelo, et al., Go-elite: a flexible solution for pathway and ontology over-representation, Bioinformatics 28 (16) (2012) 2209–2210.

[359] A. Subramanian, P. Tamayo, V.K. Mootha, S. Mukherjee, B.L. Ebert, M.A. Gillette, et al., Gene set enrichment analysis: A knowledge-based approach for interpreting genome-wide expression profiles, Proceedings of the National Academy of Sciences 102 (43) (2005) 15545–15550, https://doi.org/10.1073/pnas.0506580102, https://www.pnas.org/content/102/43/15545, https://www.pnas.org/content/102/43/15545.full.pdf.

[360] M.A.H. Ibrahim, S. Jassim, M.A. Cawthorne, K. Langlands, A topology-based score for pathway enrichment, Journal of Computational Biology 19 (5) (2012) 563–573.

[361] H. Ogata, S. Goto, W. Fujibuchi, M. Kanehisa, Computation with the KEGG pathway database, Biosystems 47 (1–2) (1998) 119–128.

[362] S. Rahmati, M. Abovsky, C. Pastrello, M. Kotlyar, R. Lu, C.A. Cumbaa, et al., pathDIP 4: an extended pathway annotations and enrichment analysis resource for human, model organisms and domesticated species, Nucleic Acids Research 48 (D1) (2019) D479–D488, https://doi.org/10.1093/nar/gkz989, https://academic.oup.com/nar/article-pdf/48/D1/D479/31697701/gkz989.pdf.

[363] I. Kuperstein, E. Bonnet, H.A. Nguyen, D. Cohen, E. Viara, L. Grieco, et al., Atlas of cancer signalling network: a systems biology resource for integrative analysis of cancer data with Google Maps, Oncogenesis 4 (7) (2015) e160.

[364] D. Nishimura, BioCarta, Biotech Software & Internet Report 2 (3) (2001) 117–120, https://doi.org/10.1089/152791601750294344.

[365] H. Ma, A. Sorokin, A. Mazein, A. Selkov, E. Selkov, O. Demin, et al., The Edinburgh human metabolic network reconstruction and its functional analysis, Molecular Systems Biology 3 (1) (2007) 135.

[366] M. Trupp, T. Altman, C.A. Fulcher, R. Caspi, M. Krummenacker, S. Paley, et al., Beyond the genome (BTG) is a (PGDB) pathway genome database: HumanCyc, Genome Biology 11 (2010) O12, https://doi.org/10.1186/gb-2010-11-s1-o12.

[367] S. Yamamoto, N. Sakai, H. Nakamura, H. Fukagawa, K. Fukuda, T. Takagi, INOH: ontology-based highly structured database of signal transduction pathways, Database 2011 (2011) Bar052, https://doi.org/10.1093/database/bar052, https://academic.oup.com/database/article-pdf/doi/10.1093/database/bar052/1272350/bar052.pdf.

[368] P.K. Sreenivasaiah, S. Rani, J. Cayetano, N. Arul, D.H. Kim, IPAVS: Integrated pathway resources, analysis and visualization system, Nucleic Acids Research 40 (D1) (2012) D803–D808.

[369] K. Kandasamy, S.S. Mohan, R. Raju, S. Keerthikumar, G.S.S. Kumar, A.K. Venugopal, et al., NetPath: a public resource of curated signal transduction pathways, Genome Biology 11 (1) (2010) 1–9.

[370] É.M. Simão, H.B. Cabral, M.A. Castro, M. Sinigaglia, J.C. Mombach, G.R. Librelotto, Modeling the human genome maintenance network, Physica A: Statistical Mechanics and its Applications 389 (19) (2010) 4188–4194.

[371] H. Mi, B. Lazareva-Ulitsky, R. Loo, A. Kejariwal, J. Vandergriff, S. Rabkin, et al., The panther database of protein families, subfamilies, functions and pathways, Nucleic Acids Research 33 (suppl_1) (2005) D284–D288.

[372] M. Whirl-Carrillo, E.M. McDonagh, J. Hebert, L. Gong, K. Sangkuhl, C. Thorn, et al., Pharmacogenomics knowledge for personalized medicine, Clinical Pharmacology & Therapeutics 92 (4) (2012) 414–417.

[373] C.F. Schaefer, K. Anthony, S. Krupa, J. Buchoff, M. Day, T. Hannay, et al., PID: the pathway interaction database, Nucleic Acids Research 37 (suppl_1) (2009) D674–D679.

[374] L. Calzone, A. Gelay, A. Zinovyev, F. Radvanyi, E. Barillot, A comprehensive modular map of molecular interactions in RB/E2F pathway, Molecular Systems Biology 4 (1) (2008) 0174.

[375] G. Joshi-Tope, M. Gillespie, I. Vastrik, P. D'Eustachio, E. Schmidt, B. de Bono, et al., Reactome: a knowledgebase of biological pathways, Nucleic Acids Research 33 (suppl_1) (2005) D428–D432.

[376] D. Fazekas, M. Koltai, D. Türei, D. Módos, M. Pálfy, Z. Dúl, et al., SignaLink 2–a signaling pathway resource with multi-layered regulatory networks, BMC Systems Biology 7 (1) (2013) 1–15.

[377] L. Perfetto, L. Briganti, A. Calderone, A. Cerquone Perpetuini, M. Iannuccelli, F. Langone, et al., Signor: a database of causal relationships between biological entities, Nucleic Acids Research 44 (D1) (2016) D548–D554.

[378] T. Jewison, Y. Su, F.M. Disfany, Y. Liang, C. Knox, A. Maciejewski, et al., SMPDB 2.0: big improvements to the small molecule pathway database, Nucleic Acids Research 42 (D1) (2014) D478–D484.

[379] A. Paz, Z. Brownstein, Y. Ber, S. Bialik, E. David, D. Sagir, et al., Spike: a database of highly curated human signaling pathways, Nucleic Acids Research 39 (suppl_1) (2011) D793–D799.

[380] N.R. Gough, Science's signal transduction knowledge environment: the connections maps database, Annals of the New York Academy of Sciences 971 (1) (2002) 585–587.

[381] H. Kitano, A. Funahashi, Y. Matsuoka, K. Oda, Using process diagrams for the graphical representation of biological networks, Nature Biotechnology 23 (8) (2005) 961–966.

[382] U. Raudvere, L. Kolberg, I. Kuzmin, T. Arak, P. Adler, H. Peterson, et al., g:Profiler: a web server for functional enrichment analysis and conversions of gene lists (2019 update), Nucleic Acids Research 47 (W1) (2019) W191–W198, https://doi.org/10.1093/nar/gkz369, https://academic.oup.com/nar/article-pdf/47/W1/W191/28879887/gkz369.pdf.

[383] B. Dutta, A. Wallqvist, J. Reifman, PathNet: a tool for pathway analysis using topological information, Source Code for Biology and Medicine 7 (1) (2012) 10, https://doi.org/10.1186/1751-0473-7-10.

[384] Q. Yang, S. Wang, E. Dai, S. Zhou, D. Liu, H. Liu, et al., Pathway enrichment analysis approach based on topological structure and updated annotation of pathway, Briefings in Bioinformatics 20 (1) (2019) 168–177.

[385] R. Feldman, I. Dagan, Knowledge discovery in textual databases (KDT), in: KDD, vol. 95, 1995, pp. 112–117.

[386] M. Allahyari, S. Pouriyeh, M. Assefi, S. Safaei, E.D. Trippe, J.B. Gutierrez, et al., A brief survey of text mining: Classification, clustering and extraction techniques, arXiv preprint, arXiv:1707.02919, 2017.

[387] R. Feldman, J. Sanger, The Text Mining Handbook: Advanced Approaches in Analyzing Unstructured Data, Cambridge University Press, 2007.

[388] X. Qiu, T. Sun, Y. Xu, Y. Shao, N. Dai, X. Huang, Pre-trained models for natural language processing: A survey, Science China. Technological Sciences (2020) 1–26.

[389] A. Vaswani, N. Shazeer, N. Parmar, J. Uszkoreit, L. Jones, A.N. Gomez, et al., Attention is all you need, in: Advances in Neural Information Processing Systems, 2017, pp. 5998–6008.

[390] J. Devlin, M.W. Chang, K. Lee, K. Toutanova, BERT: Pre-training of deep bidirectional transformers for language understanding, in: Proceedings of the 2019 Conference of the North American Chapter of the Association for Computational Linguistics: Human Language Technologies, Volume 1 (Long and Short Papers), Association for Computational Linguistics, Minneapolis, Minnesota, 2019, pp. 4171–4186, https://doi.org/10.18653/v1/N19-1423, https://aclanthology.org/N19-1423.
[391] J. Lee, W. Yoon, S. Kim, D. Kim, S. Kim, C.H. So, et al., BioBERT: a pre-trained biomedical language representation model for biomedical text mining, Bioinformatics 36 (4) (2020) 1234–1240.
[392] K. Huang, J. Altosaar, R. Ranganath, ClinicalBERT: Modeling clinical notes and predicting hospital readmission, arXiv preprint, arXiv:1904.05342, 2019.
[393] L. Rasmy, Y. Xiang, Z. Xie, C. Tao, D. Zhi, Med-BERT: pretrained contextualized embeddings on large-scale structured electronic health records for disease prediction, npj Digital Medicine 4 (1) (2021) 1–13.
[394] Y. Gu, R. Tinn, H. Cheng, M. Lucas, N. Usuyama, X. Liu, et al., Domain-specific language model pretraining for biomedical natural language processing, arXiv:2007.15779, 2020.
[395] K.S. Kalyan, A. Rajasekharan, S. Sangeetha, AMMU: A survey of transformer-based biomedical pretrained language models, Journal of Biomedical Informatics 126 (2022) 103982, https://doi.org/10.1016/j.jbi.2021.103982, https://www.sciencedirect.com/science/article/pii/S1532046421003117.
[396] B. De Bruijn, J. Martin, Getting to the (c)ore of knowledge: mining biomedical literature, International Journal of Medical Informatics 67 (1–3) (2002) 7–18.
[397] D.M. Blei, A.Y. Ng, M.I. Jordan, Latent Dirichlet allocation, Journal of Machine Learning Research 3 (2003) 993–1022.
[398] B. Liu, Sentiment analysis and opinion mining, Synthesis Lectures on Human Language Technologies 5 (1) (2012) 1–167.
[399] C. Zucco, B. Calabrese, M. Cannataro, Sentiment analysis and affective computing for depression monitoring, in: 2017 IEEE International Conference on Bioinformatics and Biomedicine (BIBM), IEEE, 2017, pp. 1988–1995.
[400] V. Carchiolo, A. Longheu, M. Malgeri, Using Twitter data and sentiment analysis to study diseases dynamics, in: Information Technology in Bio- and Medical Informatics, Springer, 2015, pp. 16–24.
[401] B. O'Dea, S. Wan, P.J. Batterham, A.L. Calear, C. Paris, H. Christensen, Detecting suicidality on Twitter, Internet Interventions 2 (2) (2015) 183–188.
[402] J.C. Eichstaedt, H.A. Schwartz, M.L. Kern, G. Park, D.R. Labarthe, R.M. Merchant, et al., Psychological language on Twitter predicts county-level heart disease mortality, Psychological Science 26 (2) (2015) 159–169.
[403] H.J. Kim, S.B. Park, G.S. Jo, Affective social network—happiness inducing social media platform, Multimedia Tools and Applications 68 (2) (2014) 355–374.
[404] B. Hesper, P. Hogeweg, Bioinformatica: een werkconcept, Kameleon 1 (6) (1970) 28–29.
[405] B. Hesper, P. Hogeweg, Bio-informatics: a working concept. A translation of "bioinformatica: een werkconcept" by B. Hesper and P. Hogeweg, arXiv:2111.11832, 2021.
[406] N.M. Luscombe, D. Greenbaum, M. Gerstein, What is bioinformatics? A proposed definition and overview of the field, Methods of Information in Medicine 40 (4) (2001) 346–358.
[407] A. Bayat, Science, medicine, and the future: Bioinformatics, BMJ 324 (7344) (2002) 1018–1022.

[408] W.Z. Khan, E. Ahmed, S. Hakak, I. Yaqoob, A. Ahmed, Edge computing: A survey, Future Generations Computer Systems 97 (2019) 219–235, https://doi.org/10.1016/j.future.2019.02.050.

[409] M.G.S. Murshed, C. Murphy, D. Hou, N. Khan, G. Ananthanarayanan, F. Hussain, Machine learning at the network edge: A survey, ACM Computing Surveys 54 (8) (2022) 170:1–170:37, https://doi.org/10.1145/3469029.

[410] G.M. Weber, C. Hong, N.P. Palmer, P. Avillach, S.N. Murphy, A. Gutiérrez-Sacristán, et al., International comparisons of harmonized laboratory value trajectories to predict severe COVID-19: Leveraging the 4CE collaborative across 342 hospitals and 6 countries: A retrospective cohort study, medRxiv 2021, 2021 Feb 5, https://doi.org/10.1101/2020.12.16.20247684, https://pubmed.ncbi.nlm.nih.gov/33564777/.

[411] I.S. Kohane, B.J. Aronow, P. Avillach, B.K. Beaulieu-Jones, R. Bellazzi, R.L. Bradford, et al., What every reader should know about studies using electronic health record data but may be afraid to ask, Journal of Medical Internet Research 23 (3) (2021) e22219, https://doi.org/10.2196/22219, https://www.jmir.org/2021/3/e22219.

[412] D. Dua, C. Graff, UCI machine learning repository, http://archive.ics.uci.edu/ml, 2017.

[413] J.N. Weinstein, E.A. Collisson, G.B. Mills, K.R.M. Shaw, B.A. Ozenberger, K. Ellrott, et al., The cancer genome atlas pan-cancer analysis project, Nature Genetics 45 (10) (2013) 1113–1120.

[414] H. Li, B. Handsaker, A. Wysoker, T. Fennell, J. Ruan, N. Homer, et al., The sequence alignment/map format and SAMtools, Bioinformatics 25 (16) (2009) 2078–2079.

Index

D

T

U

Printed in the United States
by Baker & Taylor Publisher Services